CONTEMPORARY'S

FOUNDATIONS

MATHEMATICS

SUSAN ECHAORE-YOON

Project Editor
Kathy Osmus

CONTEMPORARY
BOOKS
CHICAGO

Library of Congress Cataloging-in-Publication Data

Echaore-Yoon, Susan, 1953–
  Mathematics / Susan Echaore-Yoon ; project editor, Kathy Osmus.
    p.   cm. — (Contemporary's foundations)
  ISBN 0-8092-3830-6 (pbk.)
  1. Arithmetic.   I. Osmus, Kathy.   II. Title.   III. Series.
QA107.E24   1993
513—dc20   93-20197
                                                                  CIP

Photo Credits: pp. 1, 74, 98—UPI/Bettmann; pp. 119, 150—Betty Tsamis

Published by Contemporary Books, Inc.
Two Prudential Plaza, Chicago, Illinois 60601-6790
Manufactured in the United States of America
International Standard Book Number: 0-8092-3830-6

10   9   8   7   6

Editorial Director
Caren Van Slyke

Editorial
Christine Benton
Craig Bolt
Lisa Black
Eunice Hoshizaki
Leah Mayes

Editorial Production Manager
Norma Fioretti

Production Editor
Jean Farley Brown

Cover Design
Georgene Sainati

Illustrator
Clifford Hayes

Photo Research
Betty Tsamis

Typography
J•B Typesetting
St. Charles, Illinois

Cover photograph © The Image Bank
Inset photograph by C. C. Cain
Photo manipulation by Waselle Graphics

CONTENTS

INTRODUCTION

Welcome to Contemporary's *Foundations: Mathematics*. With this book, you will build the computation and problem-solving skills you need to understand basic mathematics. You will learn estimation and calculator skills, and you'll see how math is used in our everyday lives.

This book is divided into the following five areas:

▶ **Whole Numbers**
▶ **Money**
▶ **Decimals**
▶ **Common Fractions**
▶ **Ratios and Percents**

The math skills you'll be learning with this book include applying the four basic operations to whole numbers, decimals, and fractions. You'll also learn how to solve problems involving money, percents, ratio, and proportion.

Foundations: Mathematics has special features that will help you build your math problem-solving skills. Keep your eye out for these:

Talk Math activities give you the chance to discuss real-life uses for the new skills being presented.

On Your Calculator activities show you step-by-step how to input problems like those in that section.

Math Notes feature important tips that will help build a deeper understanding and awareness of the featured skills.

Points to Remember boxes at the end of each chapter help you review the most important points you cover in that chapter.

A **Post-Test** at the end of the book will help you to see how well you've mastered the material in the book. The **Post-Test Answer Key** on page 172 will help you to evaluate your answers. By filling out the **Post-Test Evaluation Chart** on page 171, you will see what skills you need to review.

We hope you enjoy the interesting topics in *Foundations: Mathematics*. We also invite you to explore the other books in Contemporary's *Foundations* series: *Reading*, *Social Studies*, *Writing*, and *Science*. We wish you the best of luck with your studies.

The Editors

UNIT 1
WHOLE NUMBERS

IN THIS UNIT, YOU WILL LEARN BASIC CONCEPTS AND PROBLEM-SOLVING SKILLS. USING WHOLE NUMBERS, YOU'LL LEARN HOW TO

► identify number places and their values

► add to find the total of similar amounts

► subtract to find the difference between two similar amounts

► multiply to find the total of equal groups

► divide to find the number of equal groups or the amount in each group

CHAPTER 1 | NUMBER POWER

Count to ten.

Most likely, you said: "1, 2, 3, 4, 5, 6, 7, 8, 9, 10." They are our first ten counting numbers. They are also called **whole numbers**.

Whole numbers are part of the **decimal number system**, a way of naming numbers. Fractions and mixed numbers are also part of the system.

As you've shown, a whole number is followed by another whole number. Whole numbers never end; they go on forever. The first whole number is 0 (zero).

Besides counting, we can use whole numbers to

▶ rank a group of people or things (that is, put them in order: first, second, third, and so on)

▶ identify places, things, and people, such as addresses and licenses

▶ talk about amounts

Understanding what numbers are about can help you solve math problems. In this chapter, we will study what makes up a whole number and how it gets its **value** or worth.

TALK MATH

Do these activities with a partner or group.

1. Make a list of items in your wallet or purse that have numbers on them. What do the numbers represent (for example, an account or an address)?

2. With a partner, take turns counting people or things in the class.

Whole Number Places

Write a whole number: _____

How many **number places** does it have?

All numbers have one or more number places. Each number place is filled with a symbol called a **digit**. Ten digits are used in our system to fill number places. They are *1, 2, 3, 4, 5, 6, 7, 8, 9,* and *0.*

For example, 12 has two number places. 412 has three number places.

Number Place Names

Every number place has a name. The chart below shows the names of the first ten whole number places.

1,	2	3	4,	5	6	7,	8	9	0
↑	↑	↑	↑	↑	↑	↑	↑	↑	↑
Billions	Hundred Millions	Ten Millions	Millions	Hundred Thousands	Ten Thousands	Thousands	Hundreds	Tens	Ones

The largest number place shown above is *billions.* The digit 1 fills the *billions* place. What number place does the digit 4 fill? 9? 2? 0? 8? 3? 5?

Exercise 1

Read each whole number. What digit fills each number place? The first one is started for you.

1. 16   **a.** *tens* __1__   **b.** *ones* _____

2. 8   **a.** *ones* _____

3. 27   **a.** *tens* _____   **b.** *ones* _____

4. 104   **a.** *tens* _____   **b.** *hundreds* _____   **c.** *ones* _____

5. 436   **a.** *ones* _____   **b.** *tens* _____   **c.** *hundreds* _____

6. 591   **a.** *hundreds* _____   **b.** *tens* _____   **c.** *ones* _____

7. 1,639   **a.** *ones* _____   **b.** *hundreds* _____   **c.** *tens* _____   **d.** *thousands* _____

Check your answers on page 173.

Place Values

The name of a number place also tells us its **place value**. For example, the value of the *ones* place is 1. The chart below represents the place value for the first four whole number places.

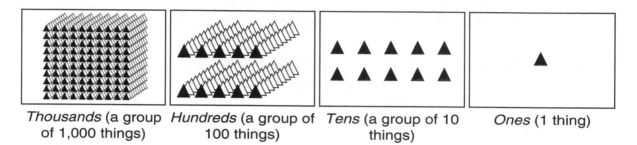

Thousands (a group of 1,000 things)   *Hundreds* (a group of 100 things)   *Tens* (a group of 10 things)   *Ones* (1 thing)

Did you notice that the values get larger as you move from right to left? The value of *ones* is shown as 1 shape (▲). The value of *tens* is shown as 10 shapes. How many shapes represent the value of the *hundreds* place (each ▲\\\\\\ equals 10)?

100 shapes represent the value of the *hundreds* place.

Exercise 2

Answer the questions about whole number place values.

1. What whole number place has the least value? _____

2. Which has the greater value: *ones* or *tens*? _____

3. Which has the greater value: *tens* or *hundreds*? _____

4. Which has the lesser value: *thousands* or *hundreds*? _____

Check your answers on page 173.

A Digit's Value

When a digit fills a number place, it takes the value of that place. For instance, the digit 4 in 14 has the value of 4 *ones*. You might think of that as 4 shapes (▲▲▲▲) in the *ones* place. The digit 1 in 14 has the value of 1 *ten*.

Example   The value of the digit 4 in different numbers:

$$14 \qquad 42 \qquad 454$$
$$\uparrow \qquad \uparrow \qquad \uparrow \ \uparrow$$

ones   tens   hundreds ones

Exercise 3

Read each number. What is the place value of the digit in color?

1. 63 _____*tens*_____

4. 824 _____

7. 1,001 _____

2. 10 _____

5. 560 _____

8. 3,475 _____

3. 77 _____

6. 125 _____

9. 9,468 _____

Check your answers on page 173.

Equivalent Values

Write the number that comes next in each series:

    ones: 5, 6, 7, 8, 9, _____   tens: 50, 60, 70, 80, 90, _____

You should have written **10** and **100**. Notice that the last number in each series has one more number place. That's the way the decimal number system works. It takes 10 of a number place to have the same value as 1 of the next higher place.

The place value chart below shows the first three number places. It shows place values as **equivalent**, or having the same value. For example, 10 *ones* = 1 *ten* and 10 *tens* = 1 *hundred*.

MATH NOTE

▶ The = sign is called **equals** sign, meaning *equal to*.

---

10 *ones* = 1 *ten*. How many *tens* are 20 *ones* equal to? If you think of counting by tens, you should get **2 *tens* = 20 *ones***.

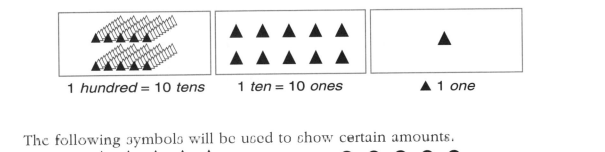

1 *hundred* = 10 tens   1 *ten* = 10 *ones*   ▲ 1 *one*

The following symbols will be used to show certain amounts.

Exercise 4

What are the equivalent values?

1. 3 *tens* = _____ *ones*

2. 5 *tens* = _____ *ones*

3. 8 *tens* = _____ *ones*

4. 12 *tens* = _____ *ones*

5. 4 *hundreds* = _____ *tens*

6. 7 *hundreds* = _____ *tens*

7. 9 *hundreds* = _____ *tens*

8. 15 *hundreds* = _____ *tens*

Check your answers on page 173.

Ones That Equal *Hundreds*

Place values that are a few places apart can be equivalent too.

Example 2 *hundreds* = 200 *ones*

Think: 200 *ones* (▲) are equal to 2 *hundreds* or 2 of these shapes (◆).

2 *hundreds*   200 *ones*

Exercise 5

What are the equivalent values?

1. 300 *ones* = _____ *hundreds*

2. 100 *ones* = _____ *hundred*

3. 400 *ones* = _____ *hundreds*

4. 600 *ones* = _____ *hundreds*

5. 5 *hundreds* = _____ *ones*

6. 8 *hundreds* = _____ *ones*

7. 3 *hundreds* = _____ *ones*

8. 9 *hundreds* = _____ *ones*

Check your answers on page 173.

How a Number Gets Its Value

When a number's place values are combined, that is the number's total value.

Example   25 = 2 *tens* and 5 *ones*

Think: 20 *ones* (▲) are equivalent to 2 *tens* or 2 of these shapes (●). 5 *ones* are left.
So 2 *tens* and 5 *ones* are equivalent to 25.

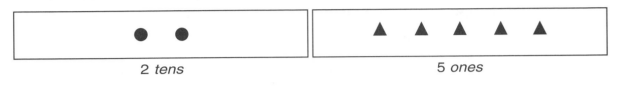

2 *tens*	5 *ones*

Exercise 6

Write the place values that combine to make up each number's value. The first one is done for you.

1. 38   =   _3 tens and 8 ones_____

2. 7   =   _____

3. 51   =   _____

4. 144   =   _____

5. 389   =   _____

6. 3,525   =   _____

Check your answers on page 173.

The Importance of Zero

The digit 0 is used as a **placeholder.** Without zeros to fill empty number places, a number cannot hold its value.

Example   60 = 6 *tens* and 0 *ones*

Think: If a zero does not fill the *ones* place in 60, the number loses its value. Without a 0 in 60, the number would have a value of 6.

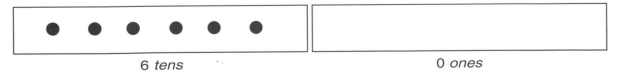

6 *tens*	0 *ones*

Regrouping

To the Next Higher Place

The digit 9 is the largest digit that can fill a number place. Sometimes, when you add or multiply numbers, you get an answer of 10 or more. When this happens, you need to **regroup** the amount. That is: change the amount so that any groups of ten are added to a *higher* place value.

Example Regroup 14 *ones* as 1 *ten* and 4 *ones*.

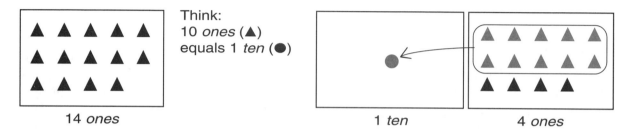

14 *ones*   1 *ten*   4 *ones*

Exercise 7

Regroup the following place values.

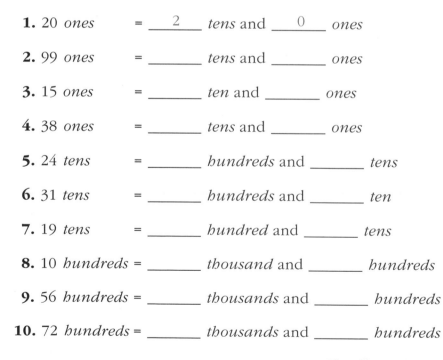

1. 20 *ones*   = ___2___ *tens* and ___0___ *ones*

2. 99 *ones*   = _____ *tens* and _____ *ones*

3. 15 *ones*   = _____ *ten* and _____ *ones*

4. 38 *ones*   = _____ *tens* and _____ *ones*

5. 24 *tens*   = _____ *hundreds* and _____ *tens*

6. 31 *tens*   = _____ *hundreds* and _____ *ten*

7. 19 *tens*   = _____ *hundred* and _____ *tens*

8. 10 *hundreds* = _____ *thousand* and _____ *hundreds*

9. 56 *hundreds* = _____ *thousands* and _____ *hundreds*

10. 72 *hundreds* = _____ *thousands* and _____ *hundreds*

Check your answers on page 173.

To the Next Lower Number Place

When you subtract or divide numbers, sometimes you need to regroup a number place. You regroup an amount so it can be added to the next *lower* number place.

Example   Regroup 2 *tens* as 20 *ones*.

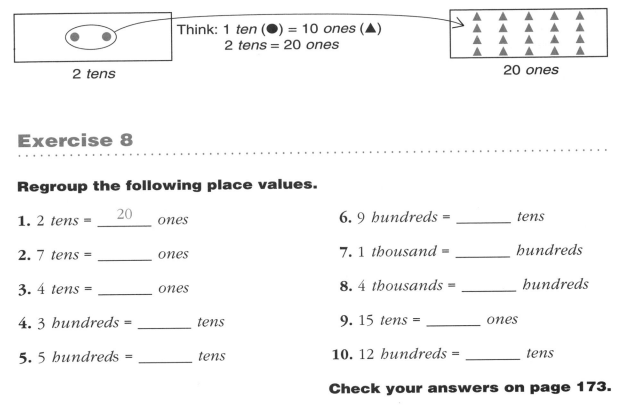

2 *tens*   Think: 1 *ten* (●) = 10 *ones* (▲)   20 *ones*
                        2 *tens* = 20 *ones*

Exercise 8

Regroup the following place values.

1. 2 *tens* = __20__ *ones*

2. 7 *tens* = _____ *ones*

3. 4 *tens* = _____ *ones*

4. 3 *hundreds* = _____ *tens*

5. 5 *hundreds* = _____ *tens*

6. 9 *hundreds* = _____ *tens*

7. 1 *thousand* = _____ *hundreds*

8. 4 *thousands* = _____ *hundreds*

9. 15 *tens* = _____ *ones*

10. 12 *hundreds* = _____ *tens*

Check your answers on page 173.

Rounded Numbers

We often use **rounded numbers** instead of exact numbers when we talk about an amount. For example, we might say we're driving about 60 miles per hour when the exact figure is 64 miles per hour.

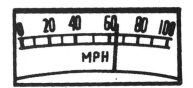

Rounded numbers should be easy and quick to work with. Many people like to use tens (10, 20, 30, etc.) and hundreds (100, 200, 300, etc.)

Here's one way to round numbers to the nearest ten: Look at the digit in the *ones* place. If the digit is less than 5, round the number down to the nearest ten. If the digit is 5 or greater, round up to the nearest ten.

Example   Round 67 and 222 to the nearest ten.

$$67 \approx 70^*$$

5 or greater, ⬆
round up

$$222 \approx 220$$

⬐ less than 5,
round down

*The $\approx$ sign means *about* or *approximately equal to*.

Exercise 9

Round each number to the nearest ten and explain why you chose to round up or down. The first one is done for you.

1. 28 $\approx$ __30__   The *ones* digit is greater than 5. _____

2. 31 $\approx$ _____   _____

3. 65 $\approx$ _____   _____

4. 121 $\approx$ _____   _____

5. 158 $\approx$ _____   _____

Check your answers on page 173.

Rounding to Hundreds

To round numbers to the nearest hundred, you can use the same rules. But now look at the digit in the *tens* place to decide if you round up or down.

Example   Round 325 and 382 to the nearest hundred.

$$325 \approx 300$$

less than 5, ⬆
round down

$$382 \approx 400$$

⬐ 5 or greater,
round up

Exercise 10

Round each number to the nearest hundred and explain why.

1. 218 $\approx$ _____   _____

2. 180 $\approx$ _____   _____

3. 452 $\approx$ _____   _____

The Calculator

Throughout this book, you can use a calculator to help you find and check answers. Below is a picture of a calculator's face. It shows the basic parts that you should find on all calculators. Read about each part and find it on your calculator. (Each type of calculator is slightly different.)

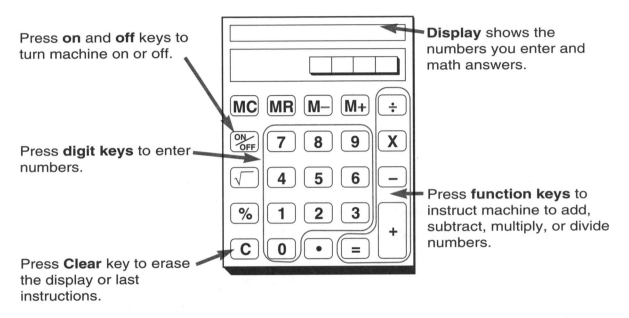

Press **on** and **off** keys to turn machine on or off.

Display shows the numbers you enter and math answers.

Press **digit keys** to enter numbers.

Press **function keys** to instruct machine to add, subtract, multiply, or divide numbers.

Press **Clear** key to erase the display or last instructions.

Each time you **input** or enter numbers on your machine for a new problem, press the **C** key first. This is like starting with a clean, blank page.

Exercise 11

Answer the questions about your calculator.

1. Turn your calculator on. What do you see on the display?

_____

2. Press the **3** key until the display is full of threes. How many number places does the display show? What's the largest possible number place?

_____

3. Press the **6** key. Then press the **C** key. What does the display show?

_____

Check your answers on page 173.

Entering Numbers

You enter numbers in the same order that you read a number. First press the digit key for the digit in a number's largest number place, then the next largest, and so on, until the number shows on your display. If you make a mistake, press the $\boxed{C}$ key and start again.

Example   Enter 190

Press digit keys $\boxed{1}$ $\boxed{9}$ $\boxed{0}$ Read display $\boxed{\qquad\qquad 190.}$

In the example, the display shows a decimal point after the *ones* place. Calculators may or may not show decimal points after whole numbers. Does yours?

Exercise 12

Practice entering these numbers into a calculator. Press the $\boxed{C}$ key after you enter each number.

1. 3	**4.** 98	**7.** 1,490	**10.** 47,090
2. 8	**5.** 201	**8.** 3,837	**11.** 842,688
3. 40	**6.** 568	**9.** 12,656	**12.** 2,696,000

Check your answers on page 173.

POINTS TO REMEMBER

▶ Whole numbers show whole amounts.

▶ All number places have a given value.

▶ Ten of one place value equals one of the next higher place. (10 *ones* = 10, 10 *tens* = 100, etc.)

▶ The first four whole number places from the greatest to the least value are *thousands*, *hundreds*, *tens*, and *ones*.

Number Checkup

How well did you understand Chapter 1?

1. Answer the questions.

    a. Why is zero an important digit?

    _____

    b. Why is 9 the largest digit that can fill a number place?

    _____

2. Combine the digits 2, 5, and 0, to make two whole numbers. Then name the number place that the digit 2 fills. **Example:** 205; 2 fills the _hundreds_ place.

    a. _____

    b. _____

3. Rank the four whole number places in order from the greatest value to the least value. Write _1_ for the greatest value, _2_ for the next greatest, and so on.

    _____ _hundreds_ _____ _ones_ _____ _tens_ _____ _thousands_

4. What numbers do the place values represent?

    a. _____ = 6 _tens_

    b. _____ = 3 _tens_ and 3 _ones_

    c. _____ = 1 _hundred_ and 0 _tens_ and 9 _ones_

5. Regroup the following amounts:

    a. 6 _tens_ as _____ _ones_

    b. 45 _tens_ as _____ _hundreds_ and _____ _tens_

6. Round the following numbers to

    a. the nearest ten: 14 ≈ _____   46 ≈ _____   388 ≈ _____

    b. the nearest hundred: 194 ≈ _____   412 ≈ _____   639 ≈ _____

7. On the digit keys below, write the order that you would enter "three hundred fifty-eight" into a calculator: ⬭ ⬭ ⬭

Check your answers on page 173.

CHAPTER 2 | ADDITION AND SUBTRACTION FACTS

In Chapters 3 and 4, we'll study problem-solving skills for addition and subtraction. But first, let's review the basic principles and math facts that the two math operations are built on.

Addition is combining groups that are alike to get a total amount.

Examples You have **2 stamps**, and you buy **3 stamps**. You now have **5 stamps**.

▶ When a number is added to 0 (zero), the answer is the number you added to 0.

You have **0 stamps**. You buy **3 stamps**. You now have **3 stamps**.

0 and ⬛⬛⬛ equals ⬛⬛⬛

▶ Numbers can be added in any order; the answer will be the same.

2 stamps and **3 stamps** are **5 stamps**;
3 stamps and **2 stamps** are **5 stamps**.

⬛⬛ and ⬛⬛⬛ equals ⬛⬛⬛⬛⬛

⬛⬛⬛ and ⬛⬛ equals ⬛⬛⬛⬛⬛

On pages 15–16 are the addition facts that you should know. They're organized in columns by their **sum**. (An answer solved by addition is called a *sum*.)

MATH NOTE

▶ The basic facts are written as **equations**. An equation shows that the numbers on both sides of the equals sign (=) are equal.

$$2 + 3 = 5 \qquad 0 + 3 = 3 \qquad 3 + 2 = 5$$

Basic Addition Facts

Example   6 + 4 = 10 ← sum

          ↑

plus sign means to add

Addition facts can be read in several ways: "Six and four equals ten. Six added to four is ten. Six plus four equals ten."

Exercise 1

Fill in the missing numbers to complete the addition facts.

Sum of 1	Sum of 2	Sum of 3	Sum of 4
1. 0 + ____ = 1	3. 0 + ____ = 2	5. 0 + ____ = 3	8. 0 + ____ = 4
2. 1 + ____ = 1	1 + 1 = 2	1 + 2 = 3	1 + 3 = 4
	4. 2 + ____ = 2	6. 2 + ____ = 3	2 + 2 = 4
		7. 3 + ____ = 3	9. 3 + ____ = 4
			10. 4 + ____ = 4

Exercise 2

Study each column of addition facts. Fill in the missing numbers.

Sum of 5	Sum of 6	Sum of 7	Sum of 8	Sum of 9
1. 0 + ____ = 5	5. 0 + ____ = 6	9. 0 + ____ = 7	14. 0 + ____ = 8	19. 0 + ____ = 9
1 + 4 = 5	1 + 5 = 6	1 + 6 = 7	1 + 7 = 8	1 + 8 = 9
2 + 3 = 5	2 + 4 = 6	2 + 5 = 7	2 + 6 = 8	2 + 7 = 9
2. 3 + ____ = 5	3 + 3 = 6	3 + 4 = 7	3 + 5 = 8	3 + 6 = 9
3. 4 + ____ = 5	6. 4 + ____ = 6	10. 4 + ____ = 7	4 + 4 = 8	4 + 5 = 9
4. 5 + ____ = 5	7. 5 + ____ = 6	11. 5 + ____ = 7	15. 5 + ____ = 8	20. 5 + ____ = 9
	8. 6 + ____ = 6	12. 6 + ____ = 7	16. 6 + ____ = 8	21. 6 + ____ = 9
		13. 7 + ____ = 7	17. 7 + ____ = 8	22. 7 + ____ = 9
			18. 8 + ____ = 8	23. 8 + ____ = 9
				24. 9 + ____ = 9

Check your answers on page 174.

Exercise 3

Fill in the missing numbers to complete each addition fact.

Sum of 10	Sum of 11	Sum of 12	Sum of 13	Sum of 14
1 + 9 = 10	2 + 9 = 11	3 + 9 = 12	4 + 9 = 13	5 + 9 = 14
2 + 8 = 10	3 + 8 = 11	4 + 8 = 12	5 + 8 = 13	6 + 8 = 14
3 + 7 = 10	4 + 7 = 11	5 + 7 = 12	6 + 7 = 13	7 + 7 = 14
4 + 6 = 10	5 + 6 = 11	6 + 6 = 12	**12.** 7 + ___ = 13	**15.** 8 + ___ = 14
5 + 5 = 10	**5.** 6 + ___ = 11	**9.** 7 + ___ = 12	**13.** 8 + ___ = 13	**16.** 9 + ___ = 14
1. 6 + ___ = 10	**6.** 7 + ___ = 11	**10.** 8 + ___ = 12	**14.** 9 + ___ = 13	
2. 7 + ___ = 10	**7.** 8 + ___ = 11	**11.** 9 + ___ = 12		
3. 8 + ___ = 10	**8.** 9 + ___ = 11			
4. 9 + ___ = 10				

Exercise 4

Study each column of addition facts. Fill in the missing numbers.

Sum of 15	Sum of 16	Sum of 17	Sum of 18
6 + 9 = 15	7 + 9 = 16	8 + 9 = 17	9 + 9 = 18
7 + 8 = 15	8 + 8 = 16	**4.** 9 + ___ = 17	
1. 8 + ___ = 15	**3.** 9 + ___ = 16		
2. 9 + ___ = 15			

Check your answers on page 174.

Basic Subtraction Facts

Subtraction is used to separate two groups that are alike or to compare amounts.

Examples You use **4** of your **5 stamps**. You have **1 stamp** left.

 take away equals

▶ When you subtract 0, the answer is the number you started with.

You have **5 stamps**. You use **0 stamps**. You still have **5 stamps** left.

take away 0 equals

▶ When a number is taken away from the same number, the answer is 0.

You use **5** of your **5 stamps**. You have **0 stamps** left.

 take away equals 0

Subtraction is the opposite of addition. Numbers that are added can be separated by subtraction. In fact, the numbers can be separated two ways.

Example   6 + 4 = 10   10 take away 4 is 6

6 and 4 are 10 → ▲ ▲ ▲ ▲ ▲ ▲ ▲ ▲ ▲ ▲

10 take away 6 is 4

For every basic addition fact, there are *two* subtraction facts. From 6 + 4 = 10, you get these subtraction facts: 10 − 6 = 4 and 10 − 4 = 6. On page 17 are the basic subtraction facts that you should know. They're organized in columns by their **difference**. (An answer found by subtraction is called a *difference*.)

10 − 6 = 4 ← difference
↑
minus sign means to subtract

You can read subtraction facts in several ways: "Ten take away six is four. Six subtracted from ten equals four. Ten minus six equals four."

POINTS TO REMEMBER

▶ Addition and subtraction are opposite math operations.

▶ An addition answer is a *sum*; a subtraction answer is a *difference*.

▶ Numbers can be added in any order, and the sum will be the same.

Exercise 5

Study each column of subtraction facts. Fill in the missing numbers.

Difference of 1	Difference of 2	Difference of 3	Difference of 4
1. 10 − ____ = 1	**11.** 11 − ____ = 2	**21.** 12 − ____ = 3	**31.** 13 − ____ = 4
2. 9 − ____ = 1	**12.** 10 − ____ = 2	**22.** 11 − ____ = 3	**32.** 12 − ____ = 4
3. 8 − ____ = 1	**13.** 9 − ____ = 2	**23.** 10 − ____ = 3	**33.** 11 − ____ = 4
4. 7 − ____ = 1	**14.** 8 − ____ = 2	**24.** 9 − ____ = 3	**34.** 10 − ____ = 4
5. 6 − ____ = 1	**15.** 7 − ____ = 2	**25.** 8 − ____ = 3	**35.** 9 − ____ = 4
6. 5 − ____ = 1	**16.** 6 − ____ = 2	**26.** 7 − ____ = 3	**36.** 8 − ____ = 4
7. 4 − ____ = 1	**17.** 5 − ____ = 2	**27.** 6 − ____ = 3	**37.** 7 − ____ = 4
8. 3 − ____ = 1	**18.** 4 − ____ = 2	**28.** 5 − ____ = 3	**38.** 6 − ____ = 4
9. 2 − ____ = 1	**19.** 3 − ____ = 2	**29.** 4 − ____ = 3	**39.** 5 − ____ = 4
10. 1 − ____ = 1	**20.** 2 − ____ = 2	**30.** 3 − ____ = 3	**40.** 4 − ____ = 4

Check your answers on page 174.

Exercise 6

Study the basic subtraction facts. Fill in the missing numbers.

Difference of 5	Difference of 6	Difference of 7	Difference of 8	Difference of 9
1. 14 − ___ = 5	**11.** 15 − ___ = 6	**21.** 16 − ___ = 7	**31.** 17 − ___ = 8	**41.** 18 − ___ = 9
2. 13 − ___ = 5	**12.** 14 − ___ = 6	**22.** 15 − ___ = 7	**32.** 16 − ___ = 8	**42.** 17 − ___ = 9
3. 12 − ___ = 5	**13.** 13 − ___ = 6	**23.** 14 − ___ = 7	**33.** 15 − ___ = 8	**43.** 16 − ___ = 9
4. 11 − ___ = 5	**14.** 12 − ___ = 6	**24.** 13 − ___ = 7	**34.** 14 − ___ = 8	**44.** 15 − ___ = 9
5. 10 − ___ = 5	**15.** 11 − ___ = 6	**25.** 12 − ___ = 7	**35.** 13 − ___ = 8	**45.** 14 − ___ = 9
6. 9 − ___ = 5	**16.** 10 − ___ = 6	**26.** 11 − ___ = 7	**36.** 12 − ___ = 8	**46.** 13 − ___ = 9
7. 8 − ___ = 5	**17.** 9 − ___ = 6	**27.** 10 − ___ = 7	**37.** 11 − ___ = 8	**47.** 12 − ___ = 9
8. 7 − ___ = 5	**18.** 8 − ___ = 6	**28.** 9 − ___ = 7	**38.** 10 − ___ = 8	**48.** 11 − ___ = 9
9. 6 − ___ = 5	**19.** 7 − ___ = 6	**29.** 8 − ___ = 7	**39.** 9 − ___ = 8	**49.** 10 − ___ = 9
10. 5 − ___ = 5	**20.** 6 − ___ = 6	**30.** 7 − ___ = 7	**40.** 8 − ___ = 8	**50.** 9 − ___ = 9

Check your answers on page 174.

Facts Checkup

How well did you understand Chapter 2?

1. Why is each answer true?

    a. $16 - 16 = 0$ _____

    _____

    b. $16 + 0 = 0 + 16$ _____

    _____

2. What is the sum?

    a. $3 + 2 =$ _____   **e.** $9 + 1 =$ _____   **i.** $9 + 9 =$ _____

    b. $6 + 1 =$ _____   **f.** $6 + 7 =$ _____   **j.** $8 + 7 =$ _____

    c. $0 + 3 =$ _____   **g.** $4 + 8 =$ _____   **k.** $6 + 9 =$ _____

    d. $4 + 4 =$ _____   **h.** $5 + 6 =$ _____   **l.** $7 + 7 =$ _____

3. What is the difference?

    a. $6 - 1 =$ _____   **e.** $10 - 5 =$ _____   **i.** $17 - 9 =$ _____

    b. $5 - 2 =$ _____   **f.** $11 - 7 =$ _____   **j.** $13 - 4 =$ _____

    c. $8 - 2 =$ _____   **g.** $13 - 9 =$ _____   **k.** $15 - 7 =$ _____

    d. $4 - 3 =$ _____   **h.** $12 - 3 =$ _____   **l.** $16 - 9 =$ _____

4. Find the sum for each basic fact. Then write the two basic subtraction facts that are based on the addition fact.

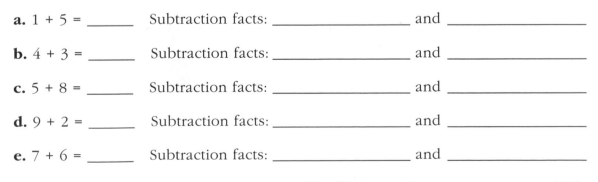

    a. $1 + 5 =$ _____   Subtraction facts: _____ and _____

    b. $4 + 3 =$ _____   Subtraction facts: _____ and _____

    c. $5 + 8 =$ _____   Subtraction facts: _____ and _____

    d. $9 + 2 =$ _____   Subtraction facts: _____ and _____

    e. $7 + 6 =$ _____   Subtraction facts: _____ and _____

Check your answers on page 174.

CHAPTER 3 | ADDITION POWER

Suppose you're making a guest list for your party. You need 9 invitations. Later, you add more names to the list. You need 7 more invitations. How many invitations do you need altogether?

The answer you need is a total amount. By combining the two amounts, you can find the answer. Count the total amount represented by the picture below.

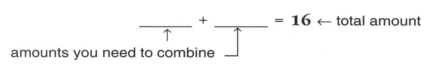

The total is **16**. You will need 16 invitations altogether.

You also can find the answer by using **addition**. Remember: addition is combining groups that are alike. Write an equation based on the pictures above.

$$\underline{\hspace{1.5cm}} + \underline{\hspace{1.5cm}} = \textbf{16} \leftarrow \text{total amount}$$

amounts you need to combine

You should have filled in **9** and **7**. These numbers are called the **addends**, the numbers to be added.

You can add several amounts as long as all the groups are *alike*. For example, you can add 1 invitation and 1 invitation. But 1 invitation and 1 name cannot be added.

In this chapter, you'll study basic concepts and skills for adding numbers. You'll practice using a calculator to get correct answers. And you'll learn about using estimation.

TALK MATH

Do these activities with a partner or group.

1. Talk about groups of objects around you that you can combine. Are the groups being added alike?

2. Describe some everyday situations in which you would need to add numbers to get the answer.

The Addition Problem

Read this situation.

A pair of children's jeans costs **$12**. A child's jacket costs **$36**. If you buy both items, what is the total cost?

To get the answer, add the numbers 12 and 36. (The sum is **$48**.)

You add the digits in the number places to get a sum. The place value chart below shows the **expression** 12 + 36. What places are being added? What is the sum of each place?

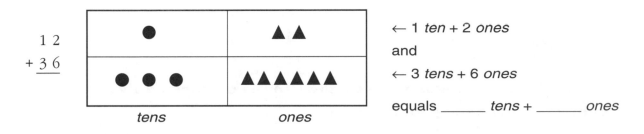

The chart shows that 1 *ten* + 3 *tens* = 4 *tens*; and 2 *ones* + 6 *ones* = 8 *ones*. The answer to $12 + $36 is $**48** (4 *tens* and 8 *ones*).

The Math Problem

You can set up a solution in two ways: a problem or an equation. The number places in a problem are lined up in straight columns. A plus sign (+) is written to show it is an addition problem. The numbers to be added can be written in any order. The *sum* will be written beneath the problem. Its number places will be lined up with those in the problem.

Example

tens ⌐ ⌐ ones		tens ⌐ ⌐ ones
1 2	OR	3 6
+ 3 6	← means equals →	+ 1 2
sum → 4 8		4 8 ← sum

Exercise 1
. . . .

Set up a math problem for each pair of numbers. Do not solve the problems.

1. 35, 18   **2.** 44, 20   **3.** 161, 465   **4.** 230, 304   **5.** 255, 99

Check your answers on page 174.

Solving the Problem

To solve an addition problem, add the place columns from right to left. As you add the digits, say their place value. Write the sum for each place column.

Example   12 + 36

Think: Add *ones*.   Add *tens*.

```
    1 2              1 2
  + 3 6            + 3 6
  ─────            ─────
      8              4 8
```
        └ Start here.

Exercise 2

Set up each pair of numbers as shown in the example. Solve the problems and check your answers.

1. 12 + 17   **3.** 15 + 15   **5.** 14 + 33   **7.** 28 + 41

2. 63 + 33   **4.** 312 + 155   **6.** 161 + 224   **8.** 433 + 433

Check your answers on page 174.

Adding Zero

When a 0 fills a number place, it is added to the other digit in its number place.

Example   4 0   Think: Add *ones*: 0 + 8 = 8 *ones*.

```
    + 2 8                   Add tens: 4 + 2 = 6 tens.
    ─────
      6 8
```

Exercise 3

Set up each pair of numbers as shown in the example. Solve the problems and check your answers.

1. 32 + 10   **3.** 40 + 30   **5.** 380 + 513

2. 20 + 25   **4.** 204 + 192   **6.** 700 + 199

Check your answers on page 174.

The Addition Equation

You can also set up an addition solution as an **equation**. Write the numbers to be added on the left side of the equals sign (=). On the right side, use a **variable** to stand for the unknown answer you're solving for. The variable can be a letter, question mark, or blank line.

Example $12 + 36 = ?$ or $36 + 12 = x$
                variable (sum) ⬏   ⬐ variable (sum)

Exercise 4

Write an equation for each pair of numbers. Do not solve the equations.

1. 33, 42   **3.** 125, 125   **5.** 416, 64

_____   _____   _____

2. 51, 18   **4.** 333, 500   **6.** 85, 909

_____   _____   _____

Check your answers on page 175.

ON YOUR CALCULATOR

Write an addition equation to help you input a problem in the correct order. Follow this procedure to input an addition equation:

Example $12 + 36 = ?$

1. Clear your calculator: $\boxed{C}$

2. Enter an addend: $\boxed{1}$ $\boxed{2}$

3. Press the plus sign: $\boxed{+}$

4. Enter an addend: $\boxed{3}$ $\boxed{6}$

5. Press the equals sign: $\boxed{=}$

6. Read the display: $\boxed{\qquad 48.}$

Practice entering the equations in exercise 4.

Carrying to the Next Higher Place

When the sum of a place column is 10 or more, you must regroup the amount and add an amount to the next higher place. (Reread page 8 to review regrouping.)

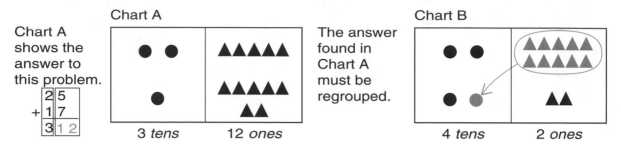

Chart A shows the answer to this problem.

$$\begin{array}{r} 2\,5 \\ +\,1\,7 \\ \hline 3\,1\,2 \end{array}$$

The answer found in Chart A must be regrouped.

Chart B shows that 10 *ones* are regrouped as 1 *ten* and carried to the *tens* place. So, 25 + 17 is **42** (4 *tens* + 2 *ones*).

Solving the Problem

When you regroup in an addition problem, show the carried (regrouped) amount. The carried amount is added with the other digits in the next higher place column.

Example   25 + 17 = _____

Think: Add *ones*.   Add *tens*.

$$\begin{array}{r} 1 \\ 2\,5 \\ +\,1\,7 \\ \hline 2 \end{array} \qquad \begin{array}{r} 1 \\ 2\,5 \\ +\,1\,7 \\ \hline 4\,2 \end{array}$$

Exercise 5

Solve the problems. Check your answers. If you get a different answer, redo the problem.

1. 15 + 27 = _____

4. 55 + 29 = _____

7. 138 + 358 = _____

2. 18 + 12 = _____

5. 48 + 47 = _____

8. 270 + 170 = _____

3. 36 + 26 = _____

6. 106 + 115 = _____

9. 356 + 291 = _____

Check your answers on page 175.

The Largest Number Place

When the sum of the largest place column is 10 or more, it also needs to be regrouped. Write the total amount in the sum.

Example   Think: Add *ones*.   Add *tens*.

```
      6 7          6 7
    + 8 2        + 8 2    Regroup: 14 tens = 1 hundred + 4 tens.
    ─────        ─────    Carry 1 hundred.
        9        1 4 9
```

Exercise 6

Solve the problems. Check your answers. If you get a different answer, redo the problem.

1. 3 6
   + 9 1
	3. 6 5
	   + 7 0
		5. 2 5
		   + 9 3
			7. 9 5
			   + 1 2
				9. 8 2
				   + 4 5

2. 4 2
   + 6 6
	4. 7 3
	   + 7 3
		6. 8 0
		   + 6 0
			8. 9 4
			   + 9 5
				10. 7 1
				   + 6 8

Check your answers on page 175.

Adding More than Two Numbers

Sometimes you'll need to add more than two numbers at a time.

When you set up your math problem, be sure the place columns are exactly straight. Here's how to add:

▶ In your mind, add the top digit and the next digit to get a sum.

▶ To that sum, add the next digit. Do this step until all the digits are added.

Example   Suppose you spend **$25** for a desk lamp, **$19** for groceries, and **$12** for dry cleaning. How much did you spend?

Think: Write the problem.   Add *ones*.   Add *tens*.

$$
\begin{array}{r}
2\,5 \\
1\,9 \\
+\ 1\,2 \\
\hline
\end{array}
$$

$$
\begin{array}{r}
1 \\
2\,5 \\
1\,9 \\
+\ 1\,2 \\
\hline
6
\end{array}
$$
Add: 5 + 9 = 14
Add: 14 + 2 = 16
Write 6 *ones*;
carry 1 *ten*.

$$
\begin{array}{r}
1 \\
2\,5 \\
1\,9 \\
+\ 1\,2 \\
\hline
5\,6
\end{array}
$$

The answer is **56**. You would have spent $56.

Exercise 7

. . . .

Solve the problems. Check your answers. If you get a different answer, redo the problem.

1.	3.	5.	7.	9.	11.
$\begin{array}{r} 1\,3 \\ 1\,4 \\ +\ 1\,2 \\ \hline \end{array}$	$\begin{array}{r} 1\,6 \\ 2\,1 \\ +\ 1\,4 \\ \hline \end{array}$	$\begin{array}{r} 2\,1 \\ 2\,6 \\ +\ 2\,5 \\ \hline \end{array}$	$\begin{array}{r} 3\,6 \\ 1\,6 \\ +\ 2\,2 \\ \hline \end{array}$	$\begin{array}{r} 1\,1 \\ 1\,4 \\ 1\,3 \\ +\ 1\,2 \\ \hline \end{array}$	$\begin{array}{r} 2\,6 \\ 2\,4 \\ 2\,8 \\ +\ 2\,2 \\ \hline \end{array}$

2.	4.	6.	8.	10.	12.
$\begin{array}{r} 1\,5 \\ 2\,0 \\ +\ 1\,4 \\ \hline \end{array}$	$\begin{array}{r} 2\,3 \\ 1\,5 \\ +\ 3\,2 \\ \hline \end{array}$	$\begin{array}{r} 2\,4 \\ 2\,5 \\ +\ 2\,5 \\ \hline \end{array}$	$\begin{array}{r} 2\,8 \\ 1\,9 \\ +\ 3\,4 \\ \hline \end{array}$	$\begin{array}{r} 2\,2 \\ 2\,0 \\ 1\,5 \\ +\ 2\,3 \\ \hline \end{array}$	$\begin{array}{r} 2\,7 \\ 1\,9 \\ 3\,2 \\ +\ 4\,0 \\ \hline \end{array}$

Check your answers on page 175.

ON YOUR CALCULATOR

Follow this procedure to enter three or more addends.

Example   25 + 19 + 12 = _____

1. Clear your calculator: `C`

2. Enter one addend: `2` `5`   Press the plus sign: `+`

3. Enter the next addend: `1` `9`   Press the plus sign: `+`

4. Enter the last addend: `1` `2`

5. Press the equals sign: `=`   Read the display: `56.`

Estimated Sums

Estimates are approximate answers. We use them when exact answers aren't necessary. We also use them to see if our answers are reasonable. One way to estimate when adding is to round the **lead digits** of the numbers to be added. The lead digit has the largest place value in a number. (Reread pages 9–10 if you need to review rounding.)

Example   19 + 72 + 188 = ?

```
  1 1
    1 9  ≈       2 0        Think:  Round 19 up to 20; round 72 down to 70;
    7 2  ≈       7 0                round 188 up to 200.
 + 1 8 8  ≈  + 2 0 0                Add: 20 + 70 + 200 = 290
  2 7 9  ≈     2 9 0  ← estimate
```

290 is a reasonable estimate.

Note: ≈ means *approximately equal to.*

Exercise 8

Estimate by rounding to the lead digits. The first one is done for you.

1.
```
    8 ≈    1 0
   28 ≈    3 0
 + 1 4 ≈ + 1 0
      ≈    5 0
```

4.
```
    5 2
      7
 + 3 7
```

7.
```
   1 8 9
   3 7 7
 + 2 1 2
```

10.
```
   1 2
   4 4
   2 6
 + 2 2
```

2.
```
   2 3
   2 4
 + 8 7
```

5.
```
    4 9
    5 4
 + 4 2 3
```

8.
```
   3 0 7
    7 8
 + 9 2 4
```

11.
```
   4 5
   2 1
   2 6
 + 4 1
```

3.
```
    9 2
   1 2 5
 + 1 0 8
```

6.
```
   2 7 5
   3 4 4
 + 2 5 7
```

9.
```
    4 5
   5 6 7
 + 3 8 1
```

12.
```
   1 0 7
    7 8
    1 5
 + 1 4 2
```

Check your answers on page 175.

Knowing When to Estimate

Using estimation to solve real-life problems isn't difficult. The most important step in estimation is deciding *when* to estimate. Sometimes an exact answer is not needed. That is the perfect time to estimate.

Example   You're ordering pizza for five people. Would you need to know exactly how much each person will eat, or would an estimated amount be close enough?
   a. estimate   **b.** exact

   In this case, you would **estimate**. You would have no way of knowing exactly how much each person would eat.

Exercise 9

Read each question. Decide whether you need an estimate or an exact amount. Explain why.

1. How many gallons of paint are needed to paint your bedroom?
   a. estimate   **b.** exact _____

2. You're ordering a cake for a party. How many people should the cake serve?
   a. estimate   **b.** exact _____

3. You're figuring out a bill for a customer. How much should you charge?
   a. estimate   **b.** exact _____

4. How much gas do you need to fill your car's gas tank?
   a. estimate   **b.** exact _____

5. You're balancing your checkbook to see if you can buy a jacket. How much is in your account?
   a. estimate   **b.** exact _____

6. You're going to a restaurant for lunch. You have to be back at work by 1 P.M. How long will it take to travel to and from the restaurant?
   a. estimate   **b.** exact _____

Check your answers on page 175.

Addition Checkup

How well did you understand Chapter 3?

1. Answer the questions.

   a. Why would you add numbers to solve a problem?

   _____

   b. Which pair of amounts could you add? Why?

    6 miles, 4 miles OR 6 gallons, 4 miles

   _____

2. Finish explaining how the problem is solved.

```
  1
  2 2 9      First, add the ones place: 9 + 4 = 13 ones.
+ 3 5 4     _____
  5 8 3     _____
```

3. Estimate the sums. (Then use a calculator or solve the problems yourself to see how close your estimates are.)

  a.   4 3   **b.** 1 1 6   **c.** 6 7   **d.** 4 5
          8   7 2   2 7   1 5
    + 5 4   + 2 0 0   + 1 9   1 5
                                       + 1 2

4. Decide whether an estimate or an exact answer is needed. Explain why.

You're on a budget. How much do you spend each month on utilities, food, and housing?

   a. estimate   **b.** exact _____

Check your answers on page 175.

POINTS TO REMEMBER

▶ A sum of 10 or more in a number place must be regrouped and an amount carried to the next higher place.

▶ When you add whole numbers, start adding with the *ones* column.

▶ In an equation, use a variable to stand for the unknown amount.

CHAPTER 4 | SUBTRACTION POWER

Imagine that you're mailing a loan payment. It's the 7th payment. You will make 15 payments in all. How many payments are left?

The answer you need is the **difference** between the two amounts. The chart below represents the amounts. Count the remainder.

You will have **8** payments left. Instead of counting, you can use *subtraction* to find the answer. Remember: subtraction is separating groups that are alike. The equation that you would solve is the following basic subtraction fact. Fill in the missing numbers.

~~1~~	~~6~~	11
~~2~~	~~7~~	12
3	8	13
~~4~~	9	14
~~5~~	10	15

15 payments take away 7 payments. Count the amount left.

$$\underline{\qquad} - \underline{\qquad} = 8 \leftarrow \text{remainder}$$
total amount ⤴   ⤴ amount that is taken away

You should have filled in **15** and **7**. You would also use subtraction for:

▶ finding the difference between two amounts

Example   A bank offers two loan schedules. Loan A has **15 payments**. Loan B has **7 payments**. How many fewer payments does Loan B have?

$$15 - 7 = 8 \leftarrow \text{difference between Loan A and Loan B}$$

▶ finding how much more is needed to reach a given amount

Example   A company offers 3 weeks of vacation after **10 years** of service. An employee has worked there for **4 years**. How many more years does the person need to work there to receive 3 weeks of vacation?

$$10 - 4 = 6 \leftarrow \text{how much more is needed}$$

TALK MATH

Do these activities with a partner or group.

1. Describe some situations in which you would subtract to get the answer.

2. Use time (minutes, hours, etc.) to show basic subtraction facts.

The Subtraction Problem

Read the situation.

A store sells twin-size comforters at **$59**. For its sale, the store lowers the price to **$25**. By how much was the price reduced?

To get the answer, subtract 25 from 59. (The difference is **34**.)

You subtract the digits in the same number place to get a difference. The place value chart below shows the expression 59 − 25. What is the difference for each place?

$$\begin{array}{r} 5\ 9 \\ -\ 2\ 5 \\ \hline \end{array}$$

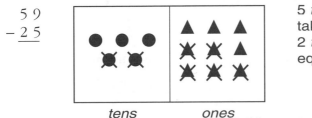

tens   ones

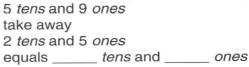

5 *tens* and 9 *ones*
take away
2 *tens* and 5 *ones*
equals _____ *tens* and _____ *ones*

The chart shows that 5 *tens* − 2 *tens* = 3 *tens*; and 9 *ones* − 5 *ones* = 4 *ones*. The answer to 59 − 25 is **34** (3 *tens* and 4 *ones*).

Solving the Problem

The numbers in a subtraction problem *always* follow a certain order. The top number is the amount to subtract from, and the bottom number is the amount being subtracted.

To solve a subtraction problem, subtract the number places from right to left. After you have subtracted a place column, write the difference directly below it.

Example   59 − 25 = ?
Think: Subtract *ones*.   Subtract *tens*.

$$\begin{array}{r} 5\ 9 \\ -\ 2\ 5 \\ \hline 4 \end{array} \qquad \begin{array}{r} 5\ 9 \\ -\ 2\ 5 \\ \hline 3\ 4 \end{array}$$

Exercise 1

Rewrite each problem as shown above. Then solve the problem.
Check your answers. If you get a different answer, redo the problem.

1. 18 − 11 =   **3.** 76 − 44 =   **5.** 249 − 25 =   **7.** 49 − 25 =

2. 36 − 12 =   **4.** 85 − 5 =   **6.** 355 − 255 =   **8.** 483 − 162 =

Check your answers on page 175.

The Subtraction Equation

If you set up a subtraction equation, write a variable to stand for the unknown difference. In an equation, write the number being subtracted *after* the minus sign.

Example $59 - 25 = ?$ ← variable (difference)
                          └ number being subtracted

Exercise 2

Write an equation for each pair of numbers. Do not solve the equations.

1. subtract 28 from 39   **3.** 56 minus 16   **5.** 500 minus 200

2. 49 minus 42   **4.** 87 subtracted from 92   **6.** 618 minus 394

Check your answers on page 175.

ON YOUR CALCULATOR

Write an equation to help you input a subtraction solution in the correct order. Follow these steps to input the solution:

Example $59 - 25 = ?$

1. Clear your calculator: | C |

2. Enter the beginning amount: | 5 | | 9 |

3. Press the minus sign: | − |

4. Enter the number to be subtracted: | 2 | | 5 |

5. Press the equals sign: | = |

6. Read the display: | 34. |

Practice entering the equations in exercise 2.

---

32 | SUBTRACTION POWER

Borrowing

Sometimes a digit in a number place is larger than the digit it is to be subtracted from. To subtract, you need to **borrow** 1 from the next higher place and regroup.

Example   21 – 14 = ?

In this equation, the *ones* place can't be subtracted without regrouping.

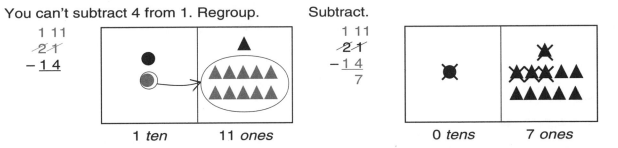

The second chart shows that 1 *ten* and 4 *ones* are subtracted from 1 *ten* and 11 *ones*. The difference is **7** (7 *ones*).

$$\begin{array}{r} 1 \\ \cancel{2}1 \\ -\,1\,4 \\ \hline \end{array}$$ Think: Subtract *ones*: Can't: 1 is smaller than 4.
Borrow 1 *ten*. That leaves: 2 – 1 = 1 *ten*.

$$\begin{array}{r} 1\ 11 \\ \cancel{2}\cancel{1} \\ -\,1\,4 \\ \hline 7 \end{array}$$ Think: Regroup 1 *ten* as 10 *ones* and add them to the *ones* place.
10 + 1 = 11 *ones*. Now subtract *ones*: 11 – 4 = 7 *ones*.
Subtract *tens*: 1 – 1 = 0 *tens*. Leave the place empty.

Exercise 3

Solve the problems. You can check your answers on a calculator. If you get a different answer, redo the problem.

1. 2 4
  – 1 6

3. 4 8
  – 3 9

5. 7 2
  – 2 5

7. 8 1
  – 1 7

9. 2 5
  –   6

11. 5 2
  –   7

2. 3 5
  – 1 9

4. 5 6
  – 2 7

6. 6 4
  – 5 5

8. 9 5
  – 4 8

10. 3 8
  –   9

12. 8 6
  –   8

Check your answers on page 175.

Subtracting Larger Number Places

Now let's look at an example of borrowing and regrouping in larger numbers.

Example   638 − 152 = ?

Write the problem.   Subtract *ones*.

$$\begin{array}{r} 6\ 3\ 8 \\ -\ 1\ 5\ 2 \\ \hline \end{array}$$

$$\begin{array}{r} 6\ 3\ 8 \\ -\ 1\ 5\ 2 \\ \hline 6 \end{array}$$

Subtract *tens*.

5 13
$$\begin{array}{r} 6\!\!\!/\,3\!\!\!/\ 8 \\ -\ 1\ 5\ 2 \\ \hline 8\ 6 \end{array}$$

Borrow 1 *hundred* and regroup as 10 *tens*. 10 *tens* + 3 *tens* = 13 *tens*

Subtract *hundreds*.

5 13
$$\begin{array}{r} 6\!\!\!/\,3\!\!\!/\ 8 \\ -\ 1\ 5\ 2 \\ \hline 4\ 8\ 6 \end{array}$$

Exercise 4

**Rewrite each problem as shown above. Then solve the problem.
Check your answer. If you get a different answer, redo the problem.**

1. 248 − 190 =
$$\begin{array}{r} 2\ 4\ 8 \\ -\ 1\ 9\ 0 \\ \hline \end{array}$$

4. 337 − 164 =

7. 625 − 584 =

10. 284 − 69 =

2. 215 − 152 =

5. 352 − 169 =

8. 655 − 618 =

11. 542 − 319 =

3. 266 − 117 =

6. 344 − 128 =

9. 739 − 99 =

12. 824 − 418 =

Check your answers on page 176.

Borrowing for a Zero

What happens when a zero fills the number place you're trying to subtract from? If that happens, borrow 1 from the next higher place.

Example   40 − 26 = ?

Think: Borrow.

3 10
$$\begin{array}{r} 4\!\!\!/\ 0\!\!\!/ \\ -\ 2\ 6 \\ \hline \end{array}$$

Borrow 1 *ten* and regroup as 10 *ones*.

Subtract *ones*.

3 10
$$\begin{array}{r} 4\!\!\!/\ 0\!\!\!/ \\ -\ 2\ 6 \\ \hline 4 \end{array}$$

Subtract *tens*.

3 10
$$\begin{array}{r} 4\!\!\!/\ 0\!\!\!/ \\ -\ 2\ 6 \\ \hline 1\ 4 \end{array}$$

Exercise 5

Solve the problems. You can check your answers on a calculator. If you get a different answer, redo the problem.

1. $\begin{array}{r} 20 \\ -12 \\ \hline \end{array}$	**3.** $\begin{array}{r} 40 \\ -17 \\ \hline \end{array}$	**5.** $\begin{array}{r} 60 \\ -36 \\ \hline \end{array}$	**7.** $\begin{array}{r} 90 \\ -25 \\ \hline \end{array}$	**9.** $\begin{array}{r} 209 \\ -\ 55 \\ \hline \end{array}$
2. $\begin{array}{r} 30 \\ -\ 9 \\ \hline \end{array}$	**4.** $\begin{array}{r} 50 \\ -24 \\ \hline \end{array}$	**6.** $\begin{array}{r} 70 \\ -53 \\ \hline \end{array}$	**8.** $\begin{array}{r} 80 \\ -25 \\ \hline \end{array}$	**10.** $\begin{array}{r} 350 \\ -\ 49 \\ \hline \end{array}$

Check your answers on page 176.

Estimated Differences

Remember: an estimate gives you an idea of whether an answer makes sense. You can find estimated differences by subtracting rounded numbers.

Example $87 - 9 = ?$

$$\begin{array}{rr} 87 \approx & 90 \\ -\ 9 \approx & -10 \\ \hline 78 & 80 \end{array} \quad \leftarrow \text{estimate}$$

Think: Round 87 up to 90; round 9 up to 10.
Subtract: $90 - 10 = 80$.

The estimate **80** shows that the answer **78** is a reasonable answer.

Exercise 6

Estimate to find out if each answer given is reasonable. If not, find the correct answer. The first one is done for you.

1. $\begin{array}{rr} 83 \approx & 80 \\ -21 \approx & -20 \\ \hline \cancel{52} & 60 \\ 62 & \end{array}$	**4.** $\begin{array}{r} 92 \\ -48 \\ \hline 44 \end{array}$	**7.** $\begin{array}{r} 412 \\ -131 \\ \hline 281 \end{array}$	**10.** $\begin{array}{r} 509 \\ -210 \\ \hline 299 \end{array}$
2. $\begin{array}{r} 53 \\ -\ 8 \\ \hline 61 \end{array}$	**5.** $\begin{array}{r} 75 \\ -15 \\ \hline 60 \end{array}$	**8.** $\begin{array}{r} 283 \\ -175 \\ \hline 18 \end{array}$	**11.** $\begin{array}{r} 782 \\ -253 \\ \hline 429 \end{array}$
3. $\begin{array}{r} 56 \\ -49 \\ \hline 17 \end{array}$	**6.** $\begin{array}{r} 896 \\ -412 \\ \hline 484 \end{array}$	**9.** $\begin{array}{r} 783 \\ -175 \\ \hline 608 \end{array}$	**12.** $\begin{array}{r} 782 \\ -653 \\ \hline 39 \end{array}$

Check your answers on page 176.

Solving Problems with a Strategy

We're always solving math problems in real life. Using a strategy, or plan, can help make it easier.

Let's see how one problem-solving strategy can help you solve problems.

Example   A slice of apple pie contains **400 calories**. A serving of applesauce contains **230 calories**. If you choose pie, how many more calories will you eat?

1. *Define the answer.* Ask yourself, "What do I need to solve for?" You may want to draw pictures to see what you need to solve.

 ← I need the difference in calories. →

400 calories   **230 calories**

2. *Define the facts.* What numbers do you need? What do you do with the numbers?

I need the difference in calories, so I should subtract.

3. *Set up the solution.* Write an equation or a math problem.

$400 - 230 = ?$ OR $\begin{array}{r} 400 \\ -\ 230 \\ \hline \end{array}$

4. *Find an estimate* to get a sense of the answer.

$\begin{array}{r} 400 \\ -\ 230 \\ \hline \end{array} \approx \begin{array}{r} 400 \\ -\ 200 \\ \hline 200 \end{array}$ Think: Round numbers to the lead digits.
      Subtract: $400 - 200 = 200$

Your answer should be *close to* 200 calories.

5. *Solve for the exact answer.* Use a calculator or solve it yourself.

$\boxed{4}\ \boxed{0}\ \boxed{0}\ \boxed{-}\ \boxed{2}\ \boxed{3}\ \boxed{0}\ \boxed{=}\ \boxed{\qquad 170.}$

The answer is **170**. You would eat 170 calories more if you chose to eat the pie.

Exercise 7

Read each situation and set up the solution. Find an estimate. Then solve for the exact answer. (Note: One problem is addition.)

1. The speed limit on the freeway is 55 miles per hour. In the city, it is 30 miles per hour. How much slower is the speed limit in the city?

Solution: _____ Estimate: _____ Answer: _____ miles per hour

2. Suppose you drive 62 miles from Angel Falls to Elton. Then you drive 25 miles to Jackson. How many total miles have you driven?

Solution: _____ Estimate: _____ Answer: _____ miles

3. Suppose you owe $22 for groceries. You pay with $30. What's your change?

Solution: _____ Estimate: _____ Answer: $ _____

4. Imagine that you have a work benefit of 28 vacation days. If you decide to take 19 days, how many vacation days remain?

Solution: _____ Estimate: _____ Answer: _____ vacation days

Check your answers on page 176.

POINTS TO REMEMBER

► Be sure the numbers in a subtraction problem are in the correct order.

► When you subtract whole numbers, start subtracting with the *ones* column.

► In an equation, be sure to use a variable to stand for the unknown difference.

► At times, you'll need to borrow and regroup an amount before you can subtract a digit.

Subtraction Checkup

How well did you understand Chapter 4?

1. Give two reasons why you would subtract amounts for an answer.

    a. _____

    b. _____

2. Finish explaining how the problem is solved.

$$\begin{array}{r} 8\ 12 \\ 6\,\cancel{9}\,\cancel{2} \\ -\ 3\ 5\ 7 \\ \hline 3\ 3\ 5 \end{array}$$

    First, subtract the *ones* place: 2 is smaller than 7, so need to regroup. ____

    _____

    _____

3. Estimate to find out if each answer is reasonable. If not, find the correct answer and write it below the problem. (You can use a calculator or solve the problems yourself.)

    a. $\begin{array}{r} 7\ 8 \\ -\ 3\ 3 \\ \hline 3\ 5 \end{array}$
    b. $\begin{array}{r} 2\ 3\ 3 \\ -\ 1\ 0\ 6 \\ \hline 1\ 2\ 7 \end{array}$
    c. $\begin{array}{r} 2\ 8\ 0 \\ -\ \ \ 4\ 5 \\ \hline 2\ 3\ 5 \end{array}$
    d. $\begin{array}{r} 4\ 1\ 2 \\ -\ \ \ 9\ 0 \\ \hline 2\ 2\ 2 \end{array}$

4. Read the situation and set up the solution. Find an estimate. Then use paper and pencil or a calculator to find the exact answer.

    A person is asking $750 for her used car. You'll buy it for $625, at most. How much less than the asking price are you willing to pay ?

    Solution: _____ Estimate: _____ Answer: $ _____

Check your answers on page 176.

CHAPTER 5 | MULTIPLICATION AND DIVISION FACTS

In Chapters 6 and 7, we will study problem-solving skills for multiplication and division. But first, we'll review the basic principles and math facts that the operations are built on.

Multiplication is combining *equal* groups to get a total.

Example   You need **2 stamps** on each of **3 letters**. You need **6 stamps** in all.

3 equal groups of 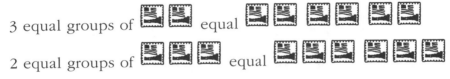 equal

▶ When a number is multiplied by 1, the answer is the number you started with.

▶ When a number is multiplied by 0 (zero), the answer is 0.

▶ Numbers can be multiplied in any order; the answer will be the same.

Example   **3** equal groups of **2 stamps** are **6 stamps**.

2 equal groups of **3 stamps** are **6 stamps**.

3 equal groups of ▨▨ equal ▨▨▨▨▨▨

2 equal groups of ▨▨▨ equal ▨▨▨▨▨▨

Division is separating a total into equal groups.

Example   **2** of you share **6 stamps** equally. So you each get **3** stamps.

▨▨▨ ▨▨▨ separated into 2 equal groups

equals ▨▨▨ per group

▶ When a number is divided by the same number, the answer is 1.

▶ When a number is divided by 1, the answer is the number you started with.

▶ When 0 is divided by a number, the answer is 0.

▶ A number **cannot** be divided by 0. In other words, an amount can never be separated into equal groups of nothing.

In the rest of this chapter, we will review basic facts for multiplication and division. The facts are the common pairs of numbers that more complex problems are based on.

Basic Multiplication Facts

Multiplication is like repeated addition of the same amount.

Example $2 \times 8 = 16$

2 equal groups of ▲▲▲▲ / ▲▲▲▲   Think: $8 + 8 = 16$

On these two pages are the basic multiplication facts that you should know. They're organized in columns by the **multiplier**. The multiplier represents how many times to combine an amount. (**Think:** How many equal groups?) Note: The multiplier can come before *or* after the multiplication sign. The **product** is the answer you get when you multiply two numbers.

Make sure you know the basic multiplication facts *before* continuing your work.

Example $2 \times 4 = 8$ ← product
          ⌐ times sign means to multiply

You can read multiplication facts in several ways: "Four multiplied by two is eight. Two times four equals eight."

Exercise 1

Study each column. Fill in the missing products.

Multiplier: 1	Multiplier: 2	Multiplier: 3	Multiplier: 4
1. $1 \times 1 =$ ___	**10.** $2 \times 1 =$ ___	**19.** $3 \times 1 =$ ___	**28.** $4 \times 1 =$ ___
2. $1 \times 2 =$ ___	**11.** $2 \times 2 =$ ___	**20.** $3 \times 2 =$ ___	**29.** $4 \times 2 =$ ___
3. $1 \times 3 =$ ___	**12.** $2 \times 3 =$ ___	**21.** $3 \times 3 =$ ___	**30.** $4 \times 3 =$ ___
4. $1 \times 4 =$ ___	**13.** $2 \times 4 =$ ___	**22.** $3 \times 4 =$ ___	**31.** $4 \times 4 =$ ___
5. $1 \times 5 =$ ___	**14.** $2 \times 5 =$ ___	**23.** $3 \times 5 =$ ___	**32.** $4 \times 5 =$ ___
6. $1 \times 6 =$ ___	**15.** $2 \times 6 =$ ___	**24.** $3 \times 6 =$ ___	**33.** $4 \times 6 =$ ___
7. $1 \times 7 =$ ___	**16.** $2 \times 7 =$ ___	**25.** $3 \times 7 =$ ___	**34.** $4 \times 7 =$ ___
8. $1 \times 8 =$ ___	**17.** $2 \times 8 =$ ___	**26.** $3 \times 8 =$ ___	**35.** $4 \times 8 =$ ___
9. $1 \times 9 =$ ___	**18.** $2 \times 9 =$ ___	**27.** $3 \times 9 =$ ___	**36.** $4 \times 9 =$ ___

Check your answers on page 176.

Multiplication Facts Table

The table shows the basic multiplication facts. The numbers being multiplied are in the far left column and along the top row. (Both are screened with color.) The **products** are the numbers in the remaining section.

How to use the table: Find the two numbers you want to multiply: one in the top row and one in the left column. Find where the row and column meet. Read the product.

For example, to solve 4 × 5, you could find 4 in the far left column and 5 in the top row. Find where the row and column meet. The answer is **20**.

×	1	2	3	4	5	6	7	8	9
1	1	2	3	4	5	6	7	8	9
2	2	4	6	8	10	12	14	16	18
3	3	6	9	12	15	18	21	24	27
4	4	8	12	16	20	24	28	32	36
5	5	10	15	20	25	30	35	40	45
6	6	12	18	24	30	36	42	48	54
7	7	14	21	28	35	42	49	56	63
8	8	16	24	32	40	48	56	64	72
9	9	18	27	36	45	54	63	72	81

Exercise 2

Fill in the missing products.

Multiplier: 5	Multiplier: 6	Multiplier: 7	Multiplier: 8	Multiplier: 9
1. 5 × 1 = ___	5. 6 × 1 = ___	10. 7 × 1 = ___	16. 8 × 1 = ___	23. 9 × 1 = ___
2. 5 × 2 = ___	6. 6 × 2 = ___	11. 7 × 2 = ___	17. 8 × 2 = ___	24. 9 × 2 = ___
3. 5 × 3 = ___	7. 6 × 3 = ___	12. 7 × 3 = ___	18. 8 × 3 = ___	25. 9 × 3 = ___
4. 5 × 4 = ___	8. 6 × 4 = ___	13. 7 × 4 = ___	19. 8 × 4 = ___	26. 9 × 4 = ___
5 × 5 = 25	9. 6 × 5 = ___	14. 7 × 5 = ___	20. 8 × 5 = ___	27. 9 × 5 = ___
5 × 6 = 30	6 × 6 = 36	15. 7 × 6 = ___	21. 8 × 6 = ___	28. 9 × 6 = ___
5 × 7 = 35	6 × 7 = 42	7 × 7 = 49	22. 8 × 7 = ___	29. 9 × 7 = ___
5 × 8 = 40	6 × 8 = 48	7 × 8 = 56	8 × 8 = 64	30. 9 × 8 = ___
5 × 9 = 45	6 × 9 = 54	7 × 9 = 63	8 × 9 = 72	9 × 9 = 81

Check your answers on page 177.

Division Facts

Division is the opposite math operation of multiplication. For every multiplication fact, there are two division facts. From $2 \times 4 = 8$, you get these division facts: $8 \div 4 = 2$ and $8 \div 2 = 4$.

Example $2 \times 4 = 8$
  8 separated into 2 equal groups makes 4 per group

  $2 \times 4 = 8 \rightarrow$ ▲ ▲ ▲ ▲ ▲ ▲ ▲ ▲

  8 separated into 4 equal groups makes 2 per group

MATH NOTE

▶ Per group means for *every* group.

On page 43 are the basic division facts. They're organized by their **divisor**. The divisor tells how many times to separate, or divide, an amount. The **dividend** (**Think**: total amount) is the number that the divisor divides into. The answer found by division is called a **quotient**.

Example $8 \div 2 = 4 \leftarrow$ quotient
  dividend ⌐↑ ↑ ↑⌐ divisor
      division sign

Division facts can be read two ways: "Two divided into eight is four. Eight divided by two equals four."

TALK MATH

Do these activities with a partner or group.

1. Discuss different times when you use multiplication and division outside of the classroom.

2. Create multiplication and division word problems using objects in the room.

Exercise 3

Fill in the missing quotients. Think: What number times the divisor equals the dividend?

Divisor: 1	Divisor: 2	Divisor: 3	Divisor: 4
1. 9 ÷ 1 = ___	10. 18 ÷ 2 = ___	19. 27 ÷ 3 = ___	28. 36 ÷ 4 = ___
2. 8 ÷ 1 = ___	11. 16 ÷ 2 = ___	20. 24 ÷ 3 = ___	29. 32 ÷ 4 = ___
3. 7 ÷ 1 = ___	12. 14 ÷ 2 = ___	21. 21 ÷ 3 = ___	30. 28 ÷ 4 = ___
4. 6 ÷ 1 = ___	13. 12 ÷ 2 = ___	22. 18 ÷ 3 = ___	31. 24 ÷ 4 = ___
5. 5 ÷ 1 = ___	14. 10 ÷ 2 = ___	23. 15 ÷ 3 = ___	32. 20 ÷ 4 = ___
6. 4 ÷ 1 = ___	15. 8 ÷ 2 = ___	24. 12 ÷ 3 = ___	33. 16 ÷ 4 = ___
7. 3 ÷ 1 = ___	16. 6 ÷ 2 = ___	25. 9 ÷ 3 = ___	34. 12 ÷ 4 = ___
8. 2 ÷ 1 = ___	17. 4 ÷ 2 = ___	26. 6 ÷ 3 = ___	35. 8 ÷ 4 = ___
9. 1 ÷ 1 = ___	18. 2 ÷ 2 = ___	27. 3 ÷ 3 = ___	36. 4 ÷ 4 = ___

Check your answers on page 177.

Exercise 4

Fill in the missing quotients.

Divisor: 5	Divisor: 6	Divisor: 7	Divisor: 8	Divisor: 9
1. 45 ÷ 5 = ___	10. 54 ÷ 6 = ___	19. 63 ÷ 7 = ___	28. 72 ÷ 8 = ___	37. 81 ÷ 9 = ___
2. 40 ÷ 5 = ___	11. 48 ÷ 6 = ___	20. 56 ÷ 7 = ___	29. 64 ÷ 8 = ___	38. 72 ÷ 9 = ___
3. 35 ÷ 5 = ___	12. 42 ÷ 6 = ___	21. 49 ÷ 7 = ___	30. 56 ÷ 8 = ___	39. 63 ÷ 9 = ___
4. 30 ÷ 5 = ___	13. 36 ÷ 6 = ___	22. 42 ÷ 7 = ___	31. 48 ÷ 8 = ___	40. 54 ÷ 9 = ___
5. 25 ÷ 5 = ___	14. 30 ÷ 6 = ___	23. 35 ÷ 7 = ___	32. 40 ÷ 8 = ___	41. 45 ÷ 9 = ___
6. 20 ÷ 5 = ___	15. 24 ÷ 6 = ___	24. 28 ÷ 7 = ___	33. 32 ÷ 8 = ___	42. 36 ÷ 9 = ___
7. 15 ÷ 5 = ___	16. 18 ÷ 6 = ___	25. 21 ÷ 7 = ___	34. 24 ÷ 8 = ___	43. 27 ÷ 9 = ___
8. 10 ÷ 5 = ___	17. 12 ÷ 6 = ___	26. 14 ÷ 7 = ___	35. 16 ÷ 8 = ___	44. 18 ÷ 9 = ___
9. 5 ÷ 5 = ___	18. 6 ÷ 6 = ___	27. 7 ÷ 7 = ___	36. 8 ÷ 8 = ___	45. 9 ÷ 9 = ___

Check your answers on page 177.

Facts Checkup

How well did you understand Chapter 5?

1. Why is each answer true?

   a. $15 \times 0 = 0$

   _____

   b. $0 \div 15 = 0$

   _____

2. Find the missing numbers.

   a. $3 \times 3 =$ _____   **f.** $6 \times 3 =$ _____   **k.** $9 \times 7 =$ _____

   b. $1 \times 5 =$ _____   **g.** $28 \div 4 =$ _____   **l.** $36 \div 6 =$ _____

   c. $48 \div 8 =$ _____   **h.** $42 \div 7 =$ _____   **m.** $9 \times 9 =$ _____

   d. $9 \times 2 =$ _____   **i.** $9 \times 3 =$ _____   **n.** $72 \div 8 =$ _____

   e. $54 \div 6 =$ _____   **j.** $5 \times 8 =$ _____   **o.** $8 \times 7 =$ _____

3. Find the product for each multiplication fact. Then write the two division facts based on the multiplication fact.

   a. $3 \times 7 =$ _____   Division facts: _____ and _____

   b. $5 \times 2 =$ _____   Division facts: _____ and _____

   c. $8 \times 4 =$ _____   Division facts: _____ and _____

   d. $3 \times 9 =$ _____   Division facts: _____ and _____

   e. $4 \times 6 =$ _____   Division facts: _____ and _____

Check your answers on page 177.

POINTS TO REMEMBER

▶ Multiplication and division are opposite math operations.

▶ A multiplication answer is called a _product_, and a division answer is called a _quotient_.

▶ Numbers can be multiplied in any order, and the product will be the same.

▶ When you multiply by 0 or divide into 0, the answer is 0.

▶ You can't divide by 0 (zero).

CHAPTER 6 | MULTIPLICATION POWER

Imagine that you're setting up a room for a meeting. You need to make 4 equal rows with 6 chairs per row. How many chairs do you need altogether?

The answer you need is a total amount. By adding the number of chairs per row four times, you can find the answer. The picture below represents what you would combine. What is the total?

6 chairs in every row
4 equal rows
6 + 6 + 6 + 6 = ?

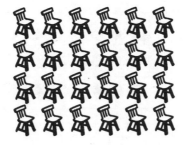

The answer is **24**. You would need a total of 24 chairs.

Another way to get the answer is to use *multiplication*. Remember: multiplication is combining equal groups to get a total. The example above could be solved using a basic multiplication fact. Fill in the missing numbers.

_____ × _____ = 24 ← total amount

number of times to ⬏
combine the amount   ⬏ amount you need to combine

You should have filled in **4** and **6**.

You use addition to combine two or more amounts that are alike.

Example   6 chairs + 4 chairs = 10 chairs

and

You use multiplication to combine the *same* amount many times.

Example   4 × 6 chairs = 24 chairs

combine 4 times

In this chapter, you'll study the basic concepts and skills for multiplying numbers. You'll practice using a calculator to get correct answers. And you'll learn how the problem-solving strategy can help you solve real-life multiplication problems.

The Multiplication Problem

Read the situation.

In a high school, **32 students** is the average number of students per class. A teacher teaches **3 classes**. About how many students will the teacher instruct in one day?

To get the answer, you multiply 32 by 3. (The product is **96 students**.)

The chart represents the answer to the expression 3×32. What places are being multiplied by the multiplier? What is the product of each place?

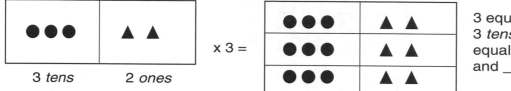 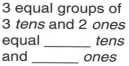

3 equal groups of
3 *tens* and 2 *ones*
equal _____ *tens*
and _____ *ones*

3 *tens*   2 *ones*

The number being multiplied is made up of *tens* and *ones*. Both places are multiplied by 3. 3×3 *tens* = 9 *tens* and 3×2 *ones* = 6 *ones*. The answer is **96** (9 *tens* and 6 *ones*).

The Math Problem

A multiplication solution can be set up as a math problem. Number places are lined up in straight columns. To show multiplication, use a times sign ($\times$).

When multiplying, the numbers can be written in any order. But if you want to keep track of what a number represents, choose a certain order. For example, you might always write the *multiplier* as the bottom number.

Example

$$\begin{array}{r} tens \searrow \swarrow ones \\ 3\,2 \\ \times \quad 3 \quad \leftarrow multiplier \\ \hline product \rightarrow \quad 9\,6 \end{array}$$

(the number of times the groups are combined)

Exercise 1

Set up a math problem for each pair of numbers. Do not solve.

1. 3 times 28

2. 45 multiplied by 6

3. 65 multiplied by 19

4. 7 times 530

5. 10 times 256

6. 709 multiplied by 12

Check your answers on page 177.

Solving the Problem

To solve a multiplication problem, you will multiply the number places from right to left, starting with the *ones* place. As you multiply digits, say their place value. Write the product for each number place directly below the problem.

Example $32 \times 3 = ?$

Think: Multiply *ones*.   Multiply *tens*.

$$
\begin{array}{r}
3\,2 \\
\times\ \ 3 \\
\hline
6
\end{array}
\qquad
\begin{array}{r}
3\,2 \\
\times\ \ 3 \\
\hline
9\,6
\end{array}
$$

Exercise 2

Rewrite each equation as a problem. Then solve the problem. Check your answer. If you get a different answer, redo the problem.

1. $3 \times 13 = ?$ **3.** $2 \times 44 = ?$ **5.** $3 \times 233 = ?$

2. $4 \times 21 = ?$ **4.** $2 \times 113 = ?$ **6.** $5 \times 111 = ?$

Check your answers on page 177.

Multiplying a Zero

When a zero fills a number place in the number being multiplied, the product for that number place is 0.

Example
$$
\begin{array}{r}
2\,0 \\
\times\ \ 3 \\
\hline
6\,0
\end{array}
$$
Think: First, multiply 3×0 *ones* $= 0$ *ones*.
Then, multiply 3×2 *tens* $= 6$ *tens*.

Exercise 3

Solve the problems.

1. $\begin{array}{r} 40 \\ \times\ 2 \\ \hline \end{array}$ **2.** $\begin{array}{r} 30 \\ \times\ 3 \\ \hline \end{array}$ **3.** $\begin{array}{r} 220 \\ \times\ 2 \\ \hline \end{array}$ **4.** $\begin{array}{r} 201 \\ \times\ 3 \\ \hline \end{array}$ **5.** $\begin{array}{r} 400 \\ \times\ 2 \\ \hline \end{array}$ **6.** $\begin{array}{r} 210 \\ \times\ 4 \\ \hline \end{array}$

Check your answers on page 177.

The Multiplication Equation

When you set up a solution as an equation, use a variable to stand for the unknown product. To keep track of what the numbers represent, choose an order that you'll always use.

Example $540 \times 2 = ?$ ← variable (product)

multiplier (pointing to 2)

Exercise 4

Write an equation for each pair of numbers. Do not solve the equations.

1. 15, multiplier: 6   **3.** 87, multiplier: 12   **5.** 249, multiplier: 5

2. 76, multiplier: 9   **4.** 56, multiplier: 10   **6.** 854, multiplier: 24

Check your answers on page 177.

ON YOUR CALCULATOR

Write an equation to help you input a multiplication solution in the correct order. Follow these steps to input the solution.

Example $32 \times 3 = ?$

1. Clear your calculator: $\boxed{C}$

2. Enter the number being multiplied: $\boxed{3}\ \boxed{2}$

3. Press the times sign: $\boxed{X}$

4. Enter the multiplier: $\boxed{3}$

5. Press the equals sign: $\boxed{=}$

6. Read the display: $\boxed{\qquad 96.}$

Practice entering the equations in exercise 4.

Regrouping in Multiplication

When the product of a number place is 10 or more, regroup the amount and carry to the next higher place. (To review regrouping, reread page 24.)

Chart A represents the answer to the expression 2 × 28. What place needs to be regrouped? How is it regrouped?

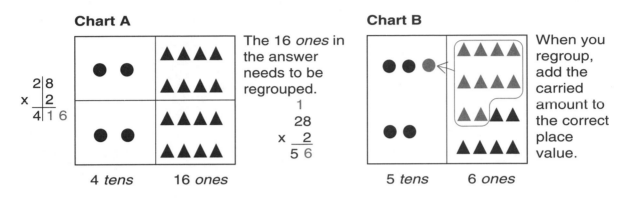

Chart A

$$\begin{array}{r} 2\,8 \\ \times\ \underline{2} \\ 4\,|\,1\,6 \end{array}$$

4 *tens*   16 *ones*

The 16 *ones* in the answer needs to be regrouped.

$$\begin{array}{r} 1 \\ 28 \\ \times\ \underline{2} \\ 5\,6 \end{array}$$

Chart B

5 *tens*   6 *ones*

When you regroup, add the carried amount to the correct place value.

Chart B shows that 16 *ones* are regrouped as 1 *ten* and 6 *ones*. The 6 *ones* are left in the *ones* place, and the 1 *ten* is carried to the *tens* place.

Multiplying and Then Adding

When you regroup and carry in multiplication, you will need to do two steps to find a product for a number place. The two steps must be done *in order*.

Step 1   For each place value, multiply the digit by the multiplier.

Step 2   Add any carried amount in that place value to the amount you found in step 1.

The example and exercise that follow will help you to practice these steps in the correct order before applying them in a multiplication problem.

Example   (2 × 2) + 1 = ?

Think: Multiply.   Add.

2 × 2 = 4   4 + 1 = 5

MATH NOTE

▶ Operations that are inside **parentheses** ( ) must be solved first.

Exercise 5

Solve the equations. You can check your answers on a calculator. If you get a different answer, redo the problem. The first one is done for you.

1. $(2 \times 2) + 5 =$ _____

     $4 \quad + 5 = 9$

4. $(3 \times 2) + 3 =$ _____

7. $(4 \times 2) + 2 =$ _____

2. $(3 \times 1) + 2 =$ _____

5. $(1 \times 6) + 1 =$ _____

8. $(2 \times 3) + 5 =$ _____

3. $(2 \times 4) + 1 =$ _____

6. $(2 \times 2) + 3 =$ _____

9. $(5 \times 0) + 3 =$ _____

Check your answers on page 177.

ON YOUR CALCULATOR

Follow these steps to check answers when you do multiplication and addition in the same problem:

Example $(2 \times 3) + 1 = ?$

1. Clear your calculator: | C |

2. Enter the multiplication problem: | 2 | | X | | 3 |

3. Press the equals sign: | = | Read display: | 6. |

4. Press the plus sign: | + |

5. Enter the amount to be added: | 1 |

6. Press the equals sign: | = |

7. Read the display: | 7. |

Solving the Problem

When you regroup in a multiplication problem, write the carried amount above the next higher place column. Add the carried amount *after* finding the product of the place.

Example $2 \times 28 = ?$

Think: Multiply *ones*.   Multiply *tens*.

$$\begin{array}{r} 1 \\ 2\,8 \\ \times\ \ 2 \\ \hline 6 \end{array}$$ Regroup as 1 *ten* and 6 *ones*.

$$\begin{array}{r} 1 \\ 2\,8 \\ \times\ \ 2 \\ \hline 5\,6 \end{array}$$ Add the carried amount *after* multiplying.

Exercise 6

Solve the problems. The first three have been started for you. You can check your answers on a calculator.

1.
$$\begin{array}{r} 1 \\ 1\,9 \\ \times\ \ 2 \\ \hline 8 \end{array}$$

4.
$$\begin{array}{r} 2\,3 \\ \times\ \ 4 \\ \hline \end{array}$$

7.
$$\begin{array}{r} 1\,1\,8 \\ \times\ \ \ \ 2 \\ \hline \end{array}$$

10.
$$\begin{array}{r} 1\,4\,3 \\ \times\ \ \ \ 3 \\ \hline \end{array}$$

2.
$$\begin{array}{r} 1 \\ 3\,5 \\ \times\ \ 2 \\ \hline 0 \end{array}$$

5.
$$\begin{array}{r} 1\,2 \\ \times\ \ 8 \\ \hline \end{array}$$

8.
$$\begin{array}{r} 2\,1\,5 \\ \times\ \ \ \ 4 \\ \hline \end{array}$$

11.
$$\begin{array}{r} 1\,8\,4 \\ \times\ \ \ \ 2 \\ \hline \end{array}$$

3.
$$\begin{array}{r} 2 \\ 2\,7 \\ \times\ \ 3 \\ \hline 1 \end{array}$$

6.
$$\begin{array}{r} 1\,6 \\ \times\ \ 6 \\ \hline \end{array}$$

9.
$$\begin{array}{r} 1\,1\,6 \\ \times\ \ \ \ 5 \\ \hline \end{array}$$

12.
$$\begin{array}{r} 1\,5\,1 \\ \times\ \ \ \ 5 \\ \hline \end{array}$$

Check your answers on page 177.

Carrying to a Zero

When the largest number place is 10 or more, regroup the amount. Write the carried amount in the product.

When a zero is in the number being multiplied, the product for that number place is 0. If there is a carried amount in that place, write the carried amount in the product.

Example $3 \times 108 = ?$

$$
\begin{array}{r}
2 \\
1\ 0\ 8 \\
\times \underline{3} \\
3\ 2\ 4
\end{array}
$$

Think: $3 \times 0 = 0$
Add the carried amount: $0 + 2 = 2$.
Write 2 in the answer.

Exercise 7

Rewrite each equation as a problem. Then solve the problem. You can check your answer on a calculator. If you get a different answer, redo the problem.

1. $4 \times 104 = ?$ **4.** $9 \times 204 = ?$ **7.** $9 \times 509 = ?$

2. $5 \times 106 = ?$ **5.** $7 \times 306 = ?$ **8.** $3 \times 409 = ?$

3. $8 \times 208 = ?$ **6.** $5 \times 305 = ?$ **9.** $6 \times 808 = ?$

Check your answers on page 177.

Multiplying by Tens

You'll often solve problems with 10 or its multiples (20, 30, etc.) as the multiplier.

When you're multiplying by 10 or one of its multiples, the product's *ones* place is always 0 (zero). Since that is the case, you can write 0 in the product. Then multiply by the *tens* place.

Example $10 \times 32 = ?$

Think: Multiply *ones*.

$$
\begin{array}{r}
3\,2 \\
\times\ 1\,0 \\
\hline
0
\end{array}
$$
$0 \times 32 = 0$
Write 0 in the *ones* place.

Multiply *tens*.

$$
\begin{array}{r}
3\,2 \\
\times\ 1\,0 \\
\hline
3\,2\,0
\end{array}
$$
1 *ten* × 2 = 2 *tens*.
1 *ten* × 3 *tens* = 3 *hundreds*.

Exercise 8

Solve the problems. You can check your answers on a calculator. If you get a different answer, redo the problem.

1. $\begin{array}{r} 1\,5 \\ \times\,1\,0 \\ \hline \end{array}$

4. $\begin{array}{r} 1\,4 \\ \times\,2\,0 \\ \hline \end{array}$

7. $\begin{array}{r} 1\,5 \\ \times\,5\,0 \\ \hline \end{array}$

10. $\begin{array}{r} 6\,4 \\ \times\,2\,0 \\ \hline \end{array}$

2. $\begin{array}{r} 2\,8 \\ \times\,1\,0 \\ \hline \end{array}$

5. $\begin{array}{r} 2\,2 \\ \times\,4\,0 \\ \hline \end{array}$

8. $\begin{array}{r} 2\,9 \\ \times\,3\,0 \\ \hline \end{array}$

11. $\begin{array}{r} 5\,2 \\ \times\,3\,0 \\ \hline \end{array}$

3. $\begin{array}{r} 3\,2 \\ \times\,1\,0 \\ \hline \end{array}$

6. $\begin{array}{r} 3\,1 \\ \times\,3\,0 \\ \hline \end{array}$

9. $\begin{array}{r} 2\,1 \\ \times\,8\,0 \\ \hline \end{array}$

12. $\begin{array}{r} 9\,9 \\ \times\,2\,0 \\ \hline \end{array}$

Check your answers on page 178.

A Two-Place Multiplier

You now know how to multiply by a one-digit number and by multiples of 10. Using that knowledge, you can solve two-digit multiplication problems.

   Think of a two-place multiplier such as 14 as two numbers, 10 and 4. Now you have two multipliers. You can make two problems out of one multiplication problem. Solve the two problems; then add the two products. That is the final product.

Example   A television shop offers a payment plan for a 27-inch TV: **$48 per month** for **14 months**. How much will the total payment be?

Think: Create two problems and multiply.
$$48 \times 14 = 48 \times 10 + 48 \times 4$$

Add products.

```
    3                              1
    4 8          4 8             1 9 2
  ×   4        × 1 0           + 4 8 0
  -----        -----           -------
  1 9 2        4 8 0             6 7 2
```

└──────── add ────────┘

Exercise 9

Finish solving each problem. Check your answers.

1. $12 \times 24 =$ _____

```
    2 4        2 4          4 8
  ×   2      × 1 0       + 2 4 0
  -----      -----
    4 8      2 4 0
```

2. $13 \times 32 =$ _____

```
    3 2        3 2
  ×   3      × 1 0
  -----      -----
    9 6      3 2 0
```

3. $15 \times 11 =$ _____

```
    1 1        1 1
  ×   5      × 1 0
  -----
    5 5
```

4. $24 \times 21 =$ _____

5. $12 \times 253 =$ _____

6. $21 \times 48 =$ _____

Check your answers on page 178.

Estimated Products

One way to estimate a product is to use rounded numbers.

Example $69 \times 2 = ?$

$$
\begin{array}{ccc}
69 & \approx & 70 \\
\times \underline{2} & & \times \underline{2} \\
138 & \approx & 140
\end{array}
$$

Think: Round 69 up to 70.
Multiply: $2 \times 70 = 140$
$\leftarrow$ estimate

Exercise 10

Estimate by rounding to the lead digit.

1. $\begin{array}{r} 43 \\ \times\ 4 \end{array}$	**3.** $\begin{array}{r} 85 \\ \times\ 9 \end{array}$	**5.** $\begin{array}{r} 112 \\ \times\ \ 3 \end{array}$	**7.** $\begin{array}{r} 185 \\ \times\ \ 4 \end{array}$
2. $\begin{array}{r} 39 \\ \times\ 6 \end{array}$	**4.** $\begin{array}{r} 56 \\ \times\ 7 \end{array}$	**6.** $\begin{array}{r} 201 \\ \times\ \ 2 \end{array}$	**8.** $\begin{array}{r} 336 \\ \times\ \ 5 \end{array}$

Check your answers on page 178.

Rounding the Multiplier

One way to estimate with a two-place multiplier is to round *both* numbers.

Example $18 \times 84 = ?$

$$
\begin{array}{ccc}
84 & \approx & 80 \\
\times \underline{18} & \approx & \times \underline{20} \\
1,512 & \approx & 1,600
\end{array}
$$

Think: Round 84 down to 80.
Round 12 down to 10.
$\leftarrow$ estimate

Exercise 11

Estimate by rounding to the lead digit.

1. $\begin{array}{r} 18 \\ \times 15 \end{array}$	**3.** $\begin{array}{r} 25 \\ \times 18 \end{array}$	**5.** $\begin{array}{r} 56 \\ \times 12 \end{array}$	**7.** $\begin{array}{r} 89 \\ \times 18 \end{array}$
2. $\begin{array}{r} 14 \\ \times 12 \end{array}$	**4.** $\begin{array}{r} 36 \\ \times 24 \end{array}$	**6.** $\begin{array}{r} 72 \\ \times 36 \end{array}$	**8.** $\begin{array}{r} 92 \\ \times 20 \end{array}$

Check your answers on page 178.

Solving Problems with Multiplication

Use multiplication to solve real-life problems if you need to combine the same amount a given number of times.

Let's see how the problem-solving strategy can help you find an answer with multiplication.

Example A train ticket from Portland to Salinas is $180. If you need 3 tickets, how much would you pay altogether?

1. *Define the answer.* Drawing pictures may help you see what needs to be solved.

   Think: "I need the cost for 3 train tickets."

2. *Define the facts.* What numbers do you need? What do you do with them?

   $180 for 1 ticket   ? for 3 tickets

   Think: "I need to combine the same amount 3 times. I can use multiplication."

3. *Set up the solution.* Write an equation or a problem.

   $3 \times 180 = $ _____   OR $\begin{array}{r} 1\,8\,0 \\ \times \quad 3 \\ \hline \end{array}$

4. *Estimate* to get a sense of the answer.

   $\begin{array}{r} 1\,8\,0 \\ \times \quad 3 \\ \hline \end{array}$ $\approx$ $\begin{array}{r} 2\,0\,0 \\ \times \quad 3 \\ \hline 6\,0\,0 \end{array}$ Think: Round to the nearest *hundred*.
   Multiply: 3 x 200 = 600
   Your answer should be *close to* $600.

5. *Solve for the exact answer.* Use a calculator or solve it yourself.

   The answer is **540.** You will need $540 for three train tickets.

Exercise 12

Read each situation and set up the solution. Estimate; then use a calculator to find the exact answer. (Note: One problem is addition.)

1. If your share of rent is $425 per month, how much will you pay for 3 months?

Solution: _____ Estimate: _____ Answer: $ _____

2. One serving of stir-fry beef contains 28 grams of protein. One gram of protein has 4 calories. How many protein calories will you eat in the serving?

Solution: _____ Estimate: _____ Answer: _____ calories

3. Suppose you pay $101 per month for car insurance. You also pay $275 per month for the car loan. How much do you pay in all for the car's loan and insurance?

Solution: _____ Estimate: _____ Answer: $ _____

4. Let's say you commute 75 miles to work each day. How many miles will you have traveled in 5 work days?

Solution: _____ Estimate: _____ Answer: _____ miles

Check your answers on page 178.

POINTS TO REMEMBER

▶ When multiplying, multiply the number places from right to left.

▶ Add a carried amount in a number place *after* finding the product for that place.

▶ When you have a two-place multiplier, think of it as two numbers. Then make two problems and solve them. Add the products for the final answer.

Multiplication Checkup

How well did you understand Chapter 6?

1. Answer the questions.

    a. How are multiplication and addition the same?

    _____

    _____

    b. How are multiplication and addition different?

    _____

    _____

2. Finish explaining how the problem is solved.

$$
\begin{array}{r}
2 \\
2\,5\,0 \\
\times 4 \\
\hline
1,0\,0\,0
\end{array}
$$

    First multiply 4×0 _ones_ = 0. _____

    _____

    _____

    _____

3. Estimate answers for the problems.

a. $\begin{array}{r}4\,1\\ \times\,2\\ \hline\end{array}$	**b.** $\begin{array}{r}5\,0\,2\\ \times\,3\\ \hline\end{array}$	**c.** $\begin{array}{r}3\,1\,7\\ \times\,3\\ \hline\end{array}$	**d.** $\begin{array}{r}3\,3\\ \times\,1\,2\\ \hline\end{array}$

4. Read the situation and set up a solution. Make an estimate. Then use a calculator or paper and pencil to find the exact answer.

A sandwich shop sells a box lunch — sandwich, chips, drink, and dessert — for $3. Your boss tells you to order 48 lunches. What will the total cost be?

Solution: _____ Estimate: _____ Answer: $ _____

Check your answers on page 178.

CHAPTER 7 | DIVISION POWER

Suppose a store has a sale on car engine oil. You and two friends buy a case of 24 quarts of oil. If you share the oil equally, how many quarts are in your share?

You need to find an equal amount per person. You want to separate the total amount into three equal groups. One way is to pass three out at a time until nothing is left. The picture below represents the amounts. How many quarts are in each group?

24 quarts of oil separated into 3 equal groups

The answer is **8**. Each person's share will be 8 quarts of oil.

Another way to get the answer is to use *division*. Remember: division is separating a total into *equal* amounts. Fill in the missing numbers using the amounts in the example above.

_____ ÷ _____ = 8   ← equal amount per group

total amount ⌐↑   ↑⌐ number of equal groups

You've seen how division can solve for the equal amount in a given number of groups. Division also solves for how many equal groups are in a total amount.

Example   Suppose you need a part-time job. You can work only **24 hours** a week. If you work **8 hours** a day, how many days will you work?

24 ÷ 8 = 3 days   ← number of equal groups

total amount ⌐↑   ↑⌐ equal amount per group

In this chapter, you will study basic concepts and skills for dividing numbers. You'll practice using a calculator to find correct answers. And you'll use the problem-solving strategy to help you solve real-life division problems.

TALK MATH

Do this activity with a partner or group.

1. Describe some situations in which you would need to divide numbers.

2. Separate things in the classroom into equal groups or find how many equal groups are in a total amount.

The Division Problem

Read this situation.

Suppose a relative's home is **84 miles** away from your home. If it takes **2 hours** to drive the distance, what's the distance you drive in 1 hour?

To get the answer, divide 84 by 2. (The quotient is **42**.)

In division, you find how many times the divisor can go evenly into each of the dividend's number places.

The chart represents the answer to the expression 84 ÷ 2. What number places are being divided? How many times does the divisor 2 go evenly into each place?

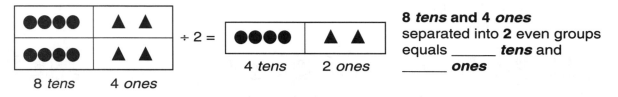

8 *tens* and 4 *ones* separated into **2** even groups equals _____ *tens* and _____ *ones*

The divisor 2 goes evenly into 8 *tens* four times, or 4 *tens*. The divisor 2 goes evenly into 4 *ones* two times, or 2 *ones*. The quotient of 84 ÷ 2 is **42** (4 *tens* and 2 *ones*).

The Math Problem

To set up the division solution as a problem, use this division sign $\overline{)}$. The *dividend* (the number to be divided) goes inside the division sign. The *divisor* (the number to divide by) goes outside the sign. The *quotient* is written above the dividend.

Example

$$2\overline{)84}$$

4 2 ← quotient
← dividend
divisor ⬏

This problem can be read as "84 divided by 2," or "2 divided into 84." The answer is **42 miles**.

Exercise 1

Set up a math problem for each pair of numbers. Do not solve.

1. 44 divided by 2   **3.** 290 divided by 4   **5.** 10 divided into 380

2. 39 divided by 6   **4.** 2 divided into 248   **6.** 96 divided by 32

Check your answers on page 178.

Solving the Problem

To solve a division problem, you also use multiplication and subtraction. The two operations help you check for the correct quotient. Divide the dividend's number places from left to right, starting with the largest place. Here's what you do:

1. *Divide* a number place by the divisor. Write the quotient above the place.

2. *Multiply* the quotient by the divisor. Write the product beneath the number place.

3. *Subtract* the product. Now bring down the digit in the next number place by writing it next to the remainder.

   Repeat the steps to divide the next number place. To help you keep track of empty places, write an **x** before you bring down a digit.

Example $2\,\overline{)\,84}$

Think: Divide *tens*.　　　Divide *ones*.

$$
\begin{array}{r}
4 \\
2\,\overline{)\,8\,4} \\
-\,8 \\
\hline
0
\end{array}
\quad
\begin{array}{l}
8 \div 2 = 4 \ tens \\
2 \times 4 = 8 \\
8 - 8 = 0
\end{array}
\qquad
\begin{array}{r}
4\,2 \\
2\,\overline{)\,8\,4} \\
-\,8\ x \\
\hline
0\,4 \\
-\,4 \\
\hline
0
\end{array}
\quad
\begin{array}{l}
4 \div 2 = 2 \\
2 \times 2 = 4 \\
4 - 4 = 0
\end{array}
$$

   You can check the quotient by multiplying it and the divisor. The answer should be the same number as the dividend. Does $2 \times 42 = 84$?

Exercise 2

Finish solving the problems. Check your answers. If you get a different answer, redo the problem. The first one is done for you.

$$
\textbf{1.}\
\begin{array}{r}
1\,3 \\
2\,\overline{)\,2\,6} \\
-\,2\ x \\
\hline
0\,6 \\
-\ \ 6 \\
\hline
0
\end{array}
\qquad
\begin{array}{r}
1\,3 \\
\times\,2 \\
\hline
2\,6
\end{array}
\ \text{Check}
$$

$$
\textbf{3.}\
\begin{array}{r}
2 \\
3\,\overline{)\,6\,3} \\
-\,6\ x \\
\hline
0\,3
\end{array}
\qquad\qquad
\textbf{5.}\
\begin{array}{r}
4 \\
2\,\overline{)\,8\,8} \\
-\,8\ x
\end{array}
\qquad\qquad
\textbf{7.}\
\begin{array}{r}
3 \\
3\,\overline{)\,9\,3}
\end{array}
$$

$$
\textbf{2.}\
\begin{array}{r}
2\,1 \\
2\,\overline{)\,4\,2} \\
-\,4\ x \\
\hline
0\,2
\end{array}
\qquad
\textbf{4.}\
\begin{array}{r}
1 \\
6\,\overline{)\,6\,6} \\
-\,6\ x \\
\hline
0\,6
\end{array}
\qquad
\textbf{6.}\
\begin{array}{r}
4 \\
2\,\overline{)\,8\,4} \\
-\,8\ x
\end{array}
\qquad
\textbf{8.}\
\begin{array}{r}
2 \\
4\,\overline{)\,8\,4}
\end{array}
$$

Check your answers on page 179.

A Division Equation

When you set up a division solution as an equation, you use this division sign ÷. The dividend is always written first — *before* the division sign. Remember to write a variable to represent the quotient you're solving for.

Example $84 ÷ 2 = x$ ← variable (quotient)
   dividend ⌐↑ ↑ ↑⌐ divisor
          divided by

Exercise 3

Set up a math equation for each pair of numbers. Do not solve the problems.

1. 50 divided by 5   **3.** 168 divided by 8   **5.** 84 divided by 12

2. 3 divided into 69   **4.** 382 divided by 2   **6.** 14 divided into 520

Check your answers on page 179.

ON YOUR CALCULATOR

Use division equations to help you enter solutions on your calculator. Follow these steps to input a division equation.

Example $84 ÷ 2 = ?$

1. Clear your calculator: [C]

2. Enter the dividend: [8] [4]

3. Press the division sign: [÷]

4. Enter the divisor: [2]

5. Press the equals sign: [=]

6. Read the display: [42.]

Practice entering the math equations in exercise 3.

Dividing Three-Place Numbers

Divide, multiply, and subtract. Remember, you do each operation to find the quotient of each number place in the dividend. After following these steps, bring down the digit of the next place to be divided. Then perform the operation steps again. Now, let's practice using the steps to divide a three-place dividend.

ONLY $396 ⁰⁰

Example   A washing machine costs **$396**. If you make **3 equal payments**, how much will each payment be?

```
        1 3 2
  3 ) 3 9 6
    − 3 x x
      0 9
      − 9
        0 6
        − 6
          0
```

Think: Divide into *hundreds*. 3 ÷ 3 = 1 *hundred*.
  3 − 3 = 0. Bring down 9 *tens*.
  Divide into *tens*. 9 ÷ 3 = 3 *tens*. 9 − 9 = 0.
  Bring down 6 *ones*.
  Divide into *ones*. 6 ÷ 3 = 2 *ones*. 6 − 6 = 0.
  Each payment will be **$132**.

Exercise 4

Solve the problems. You can check your answers on a calculator. If you get a different answer, redo the problem.

1. 4) 4 8 4   **3.** 2) 2 2 8   **5.** 2) 6 2 8   **7.** 2) 8 8 2

2. 3) 3 3 6   **4.** 5) 5 5 5   **6.** 3) 6 9 3   **8.** 3) 9 9 9

Check your answers on page 179.

Dividing into a Zero

When a dividend contains a zero, the zero is divided by the divisor. The answer is 0 and is written in the quotient. The multiplication and subtraction steps can be skipped. If another number place still needs to be divided, bring down the place's digit and write it next to the zero.

Example

$$\begin{array}{r} 1\,0\,3 \\ 2\,)\overline{2\,0\,6} \\ -\,\underline{2}\ \text{x}\ \text{x} \\ 0\,0\,6 \\ -\,\underline{6} \\ 0 \end{array}$$

Think: Divide into *hundreds*. $2 \div 2 = 1$ *hundred*. $2 - 2 = 0$.
Bring down 0 *tens*.
Divide into *tens*. $0 \div 2 = 0$. Write 0 in the quotient.
Bring down 6 *ones*.
Divide into *ones*. $6 \div 2 = 3$ *ones*. $6 - 6 = 0$.

Regrouping in Division

Sometimes the divisor is larger than a digit it's dividing into. You need to regroup the number place and add the amount to the next lower place. Then divide.

In the expression $126 \div 2$, the *hundreds* place can't be divided by 2 without regrouping. Chart A shows the amount before regrouping. How would you regroup the *hundreds* place so you can divide by 2?

Chart A

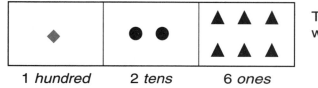

1 *hundred*   2 *tens*   6 *ones*

The *hundreds* place can't be divided by 2 without regrouping.

Chart B

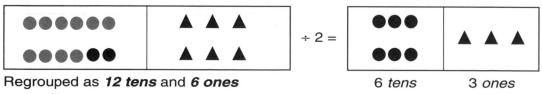

Regrouped as **12 tens** and **6 ones**

$\div 2 =$

6 *tens*   3 *ones*

Chart B shows the regrouped amount being divided: 12 *tens* and 6 *ones* are divided by the divisor 2. The divisor 2 goes evenly into 12 *tens* six times or 6 *tens*. The divisor goes evenly into 6 *ones* three times or 3 *ones*. The quotient of $126 \div 2$ is **63** (6 *tens* and 3 *ones*).

Solving the Problem

Remember: a number place needs to be regrouped when the digit is *smaller* than the divisor.

Example $126 \div 2 = ?$

Think: Divide *hundreds*.	Divide *tens*.	Divide *ones*.

$2 \overline{)126}$ 1 is smaller than 2.
Regroup 1 *hundred*
as 10 *tens*. Add to
tens place.

$$\begin{array}{r} 6 \\ 2\overline{)126} \\ -12 \\ \hline 0 \end{array}$$

$$\begin{array}{r} 63 \\ 2\overline{)126} \\ -12\text{x} \\ \hline 06 \\ -6 \\ \hline 0 \end{array}$$

Exercise 5

Solve the problems. You can check your answers on a calculator. If you get a different answer, redo the problem.

1. $3\overline{)129}$

5. $7\overline{)217}$

9. $8\overline{)408}$

2. $4\overline{)168}$

6. $4\overline{)288}$

10. $7\overline{)567}$

3. $2\overline{)184}$

7. $5\overline{)300}$

11. $6\overline{)630}$

4. $2\overline{)100}$

8. $6\overline{)366}$

12. $9\overline{)819}$

Check your answers on page 179.

Regrouping the *Tens* Place

When the divisor can't divide into the dividend's *tens* place, regroup the amount and add it to the *ones* place. To show that the *tens* place is empty, write a 0 in the quotient.

Example $615 \div 3 = ?$

Think: Divide *hundreds*.

$$\begin{array}{r} 2 \\ 3\overline{)615} \\ -\underline{6} \\ 0 \end{array}$$

Divide *tens*.

$$\begin{array}{r} 20 \\ 3\overline{)615} \\ -\underline{6}x \\ 01 \end{array}$$

1 is smaller than 3.
Regroup 1 *ten* as
10 *ones* and add
to *ones* place.

Divide *ones*.

$$\begin{array}{r} 205 \\ 3\overline{)615} \\ -\underline{6}xx \\ 015 \\ -\underline{15} \\ 0 \end{array}$$

Exercise 6

Rewrite each equation as a problem. Then solve the problem. Check your answers. If you get a different answer, redo the problem.

1. $218 \div 2 =$ _____

5. $621 \div 3 =$ _____

9. $816 \div 2 =$ _____

2. $420 \div 4 =$ _____

6. $414 \div 2 =$ _____

10. $927 \div 3 =$ _____

3. $318 \div 3 =$ _____

7. $756 \div 7 =$ _____

11. $981 \div 9 =$ _____

4. $535 \div 5 =$ _____

8. $630 \div 6 =$ _____

12. $872 \div 8 =$ _____

Check your answers on page 179.

Uneven Division

Sometimes a number place has a remainder left. You'll need to regroup the remainder and add it to the next lower place. In uneven division, any number place can have a remainder.

To solve an uneven division problem, add a regrouping step to the solution.

1. *Divide* a number place by the divisor.

2. *Multiply* the quotient by the divisor. The product should be smaller than the number place.

3. *Subtract* the product. The remainder should be smaller than the divisor.

4. *Regroup* the remainder so that it can be added to the next lower place. Write the regrouped amount. Repeat the steps if needed.

Example $56 \div 4 = ?$

Think: Divide *tens*.

$$
\begin{array}{r}
1 \\
4\overline{)5\ 6} \\
-\underline{4}\ x \\
1
\end{array}
$$

Regroup the remaining 1 *ten* as 10 *ones*.

Divide *ones*.

$$
\begin{array}{r}
1\ 4 \\
4\overline{)5\ 6} \\
-\underline{4}\ x \\
1\ 6 \\
-\underline{1\ 6} \\
0
\end{array}
$$

Add 10 *ones* to 6 *ones*. Divide 16 by 4.

Exercise 7

Finish solving each problem. You can check each answer on a calculator. If you get a different answer, redo the problem.

1.
$$
\begin{array}{r}
1 \\
2\overline{)3\ 0} \\
-\underline{2}\ x \\
1
\end{array}
$$

3. $3\overline{)5\ 4}$

5. $2\overline{)2\ 5\ 4}$

7. $6\overline{)9\ 0}$

2.
$$
\begin{array}{r}
2 \\
3\overline{)7\ 8} \\
-\underline{6}\ x \\
1
\end{array}
$$

4. $2\overline{)7\ 6}$

6. $2\overline{)4\ 3\ 6}$

8. $3\overline{)7\ 5\ 3}$

Check your answers on page 179.

A Two-Place Divisor

A two-place divisor won't divide into the first number place. You'll need to regroup the number places until you have a number larger than the divisor to divide into.

   With a two-place divisor, you often have to try several numbers before you get a correct quotient. Here's a procedure that can help you narrow the search.

1. Divide the dividend's largest number place by the divisor's *tens* digit.

2. Multiply the divisor by the answer found in step 1. Compare the product and the dividend. If the product is the same number or smaller, that's the quotient. If the product is bigger, then you need to multiply the divisor by the next smaller number.

Example 72 ÷ 12 = ?

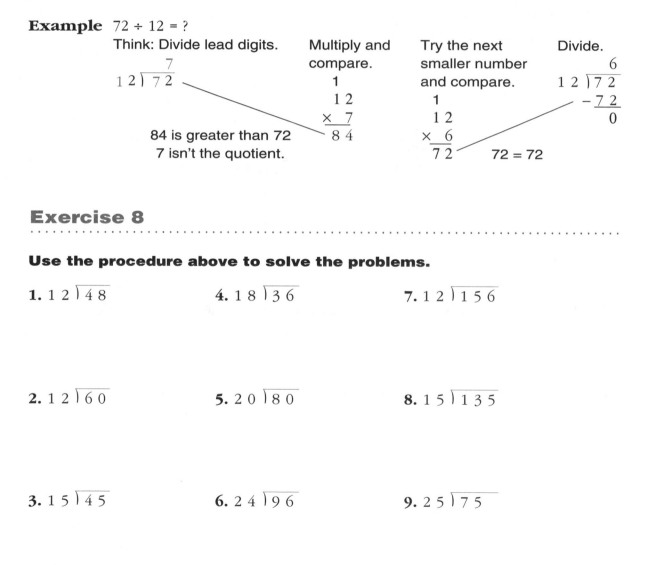

Exercise 8

Use the procedure above to solve the problems.

1. 12)48   **4.** 18)36   **7.** 12)156

2. 12)60   **5.** 20)80   **8.** 15)135

3. 15)45   **6.** 24)96   **9.** 25)75

Check your answers on page 179.

Dividing by a Two-Place Divisor

Now let's combine all the steps for dividing by a two-place divisor.

Find the lead digit in the quotient. Use the strategy you learned on page 68. Do the division steps: divide, multiply, and subtract. Regroup if necessary and find the next quotient. Repeat the same steps.

Only $ **720**

Example Suppose you buy a dining room set for **$720**. You agree to pay **$20** per month. How many monthly payments will you make?

Divide lead digits.

$$20\overline{)720} \quad 7 \div 2 \approx 3$$
$$3 \times 20 = 60$$
$$60 < 72$$

Divide *tens*.

$$\begin{array}{r} 3 \\ 20\overline{)720} \\ -60x \\ \hline 120 \end{array}$$

Regroup 12 *tens* as 120 *ones*. Add 0 *ones*.

Repeat steps.

$$\begin{array}{r} 36 \\ 20\overline{)720} \\ -60x \\ \hline 120 \\ -120 \\ \hline 0 \end{array}$$

$12 \div 2 = 6$
$6 \times 20 = 120$
$720 \div 20 = \mathbf{36}$

Exercise 9

Rewrite the equations as problems. Then solve the problems. You can check your answers on a calculator.

1. $270 \div 15 =$ _____

2. $144 \div 12 =$ _____

3. $260 \div 13 =$ _____

4. $300 \div 12 =$ _____

5. $384 \div 12 =$ _____

6. $420 \div 15 =$ _____

7. $576 \div 24 =$ _____

8. $648 \div 36 =$ _____

9. $840 \div 28 =$ _____

Check your answers on page 179.

Estimated Quotients

One way to estimate when dividing is to use **compatible numbers**. Compatible numbers are numbers that divide evenly because they are based on the basic division facts.

Here's how to use compatible numbers. Look at the divisor and dividend. Think what basic division fact you can change the problem to so it divides evenly and easily.

Example   83 ÷ 9 = ?

Think: Find compatible numbers.   Estimate.

$$9\overline{)83} \approx 9\overline{)81}$$

$$9\overline{)81}^{\,9}$$ The answer should be close to 9.

Exercise 10

Use compatible numbers to find estimates for the equations.

1. 37 ÷ 6 ≈ _____
$6\overline{)36}$

4. 68 ÷ 8 ≈ _____

7. 71 ÷ 9 ≈ _____

2. 50 ÷ 7 ≈ _____

5. 28 ÷ 3 ≈ _____

8. 31 ÷ 4 ≈ _____

3. 65 ÷ 9 ≈ _____

6. 46 ÷ 5 ≈ _____

9. 52 ÷ 8 ≈ _____

Check your answers on page 179.

Estimating with Larger Numbers

When working with larger numbers, use the first one or two digits to make a compatible pair. Then fill the other number places with zeros.

Example   135 ÷ 4 = ?

Think: Find compatible numbers.   Estimate.

$$4\overline{)135} \approx 4\overline{)120}$$
↑
fill number place

$$4\overline{)120}^{\,30}$$ The answer should be close to 30.

Exercise 11

Use compatible numbers to find estimates.

1. 57 ÷ 2 ≈ _____

2. 49 ÷ 3 ≈ _____

3. 93 ÷ 4 ≈ _____

4. 246 ÷ 6 ≈ _____

5. 302 ÷ 8 ≈ _____

6. 153 ÷ 5 ≈ _____

7. 207 ÷ 5 ≈ _____

8. 318 ÷ 4 ≈ _____

9. 486 ÷ 8 ≈ _____

Check your answers on page 180.

Estimating with Two-Digit Divisors

Sometimes you may need to change *both* the divisor and the dividend to create a compatible pair.

Example   3,990 ÷ 11 = ?

Think: Find compatible numbers.   Estimate.

$$11 \overline{)3{,}990} \approx 10 \overline{)4{,}000}$$

↑
fill number places

$$10 \overline{)4{,}000} = 400$$ The answer should
be close to 400.

Exercise 12

Estimate the quotients using compatible numbers.

1. 1,680 ÷ 39 ≈ _____

2. 1,808 ÷ 17 ≈ _____

3. 252 ÷ 14 ≈ _____

4. 1,980 ÷ 11 ≈ _____

5. 6,212 ÷ 12 ≈ _____

6. 16,589 ÷ 15 ≈ _____

7. 3,342 ÷ 12 ≈ _____

8. 6,280 ÷ 92 ≈ _____

9. 4,560 ÷ 16 ≈ _____

Check your answers on page 180.

Solving Real-Life Division Problems

Review the steps of the problem-solving strategy below. Then practice some real-life division problems.

1. *Define the answer*: What are you solving for?

2. *Define the facts*: What numbers do you need? What do you do with the numbers?

3. *Set up the solution*: Write a math equation or problem.

4. *Estimate* to get a sense of the answer.

5. *Solve* for the exact answer. Use a calculator or solve it yourself.

Exercise 13

Read each situation and set up the solution. Find an estimate and the exact answer. (Note: You'll solve one problem by subtracting.)

1. A typing test lasts 5 minutes. If you typed 310 words within the time limit, how fast do you type per minute?
Solution: _____ Estimate: _____ Answer: _____ words per minute

2. A mattress factory gives a 60-month guarantee on its products. How many years does the guarantee last? (Note: 12 months = 1 year)
Solution: _____ Estimate: _____ Answer: _____ years

3. A person lost 45 pounds when she was sick for six weeks. She weighed 156 pounds before she became ill. How much does she weigh now?
Solution: _____ Estimate: _____ Answer: _____ pounds

4. A can of vegetable soup has 324 calories. If it makes 4 servings, how many calories are in one serving?
Solution: _____ Estimate: _____ Answer: _____ calories

Check your answers on page 180.

Division Checkup

How well did you understand Chapter 7?

1. Answer the questions.

    a. How are division and subtraction alike?

    _____

    b. How are division and subtraction different?

    _____

2. Finish explaining how the problem is solved.

$$\begin{array}{r} 7\;3 \\ 2\overline{)1\,4\,6} \\ -\,1\,4\,\text{x} \\ \hline 0\;6 \\ -\;6 \\ \hline 0 \end{array}$$

1 *hundred* ÷ 2 can't be done. Regroup 1 *hundred* as _____

_____

_____

_____

3. Use compatible numbers to find an estimate for each problem. Then solve for the answer. (Use a calculator or solve the problems yourself.)

    a. $4\overline{)4\,0\,8}$ **b.** $5\overline{)4\,6\,5}$ **c.** $3\overline{)2\,2\,8}$ **d.** $12\overline{)2{,}5\,2\,0}$

4. Read the situation. Set up a solution, make an estimate, and find the exact answer.

> A car uses up 28 gallons of gasoline after 588 miles. What is the average number of miles the car can travel on one gallon of gasoline?

Solution: _____ Estimate: _____ Answer: _____ miles per gallon

Check your answers on page 180.

POINTS TO REMEMBER

▶ Be sure that the divisor and dividend are in the correct order.

▶ Divide the dividend's place values from left to right.

▶ *Divide, multiply,* and *subtract* are the three steps to dividing every place in the dividend. The fourth step, when necessary, is *regrouping.*

Unit 2
MONEY

IN THIS UNIT, YOU WILL LEARN BASIC CONCEPTS AND PROBLEM-SOLVING SKILLS WITH MONEY. YOU'LL LEARN HOW TO

► express (whole) dollars, cents, and dollar-and-cent amounts

► explain the meaning of cents

► identify cents' number places and their values

► round money to the nearest dollar, dime, and penny

► estimate answers

► use a calculator to check your answers

CHAPTER 8 | MONEY POWER

Every country has its own system of money. The American system of **money** is based on the decimal number system. **Dollars**, or whole amounts, have the same value as whole numbers. **Cents**, or fractional amounts, have the same value as **decimal fractions:** equal parts of 1.

The pictures below show some of our paper money. Each represents a whole amount. How much is each bill worth?

$ _____ $ _____ $ _____ $ _____

To show that a number represents money, a dollar sign (**$**) is written before the number. Dollar amounts can be shown with or without the number places for cents.

Example $5 = $5.00
The decimal point separates ↑ ↑ **Each 0 digit fills an empty number place.**
dollars from cents.

You can read both expressions as "Five dollars." You also can read the second expression as "Five dollars *and* no cents."

How much money is represented in the group below? Write the amount without the zeros for cents and then with the zeros. (Remember to write the $ sign before a dollar amount.)

_____ OR _____

You should have written these answers: **$100** or **$100.00**.

As you study dollar-and-cent amounts, keep in mind that you are also working with *decimals*. (You'll study more about decimals in Unit 3.)

TALK MATH

Do these activities with a partner or group.

1. Make a list of situations in which you would need to calculate with money.

2. Take turns with a partner saying a dollar amount and giving a combination of bills that add up to it. For example, you might say, "$8." Your partner could say, "A five-dollar bill and 3 one-dollar bills."

Cents

The picture shows the most common coins we use. Together, how much money do the coins make?

Altogether, the coins add up to 91¢ or 91 **cents**. (The ¢ sign stands for cents.)

Money that's less than $1 is called *cents*. Cents are in fact a fraction, or equal parts, of $1. The word cents means **hundredths** — 100 equal parts of 1. Each of our coins is worth so many *hundredths* of $1.

Let's see what *hundredths* of $1 look like. The square below represents 1. It is divided into 100 equal parts. 100¢ = $1.

Each equal part is 1 *hundredth* (Think: 1¢). 91 out of 100 equal parts are shaded (Think: 91¢).

Equal to $1

When coins total 100¢, we say "one dollar." That's because 100-*hundredths* is the same as 1.

Example   1 dime = 10¢

              10 dimes = 100¢ = $1

Exercise 1

How many of each coin equal $1?

1. _____ pennies   **2.** _____ nickels   **3.** _____ quarters   **4.** _____ 50-cent pieces

Check your answers on page 180.

Cents and Number Places

Cents can be shown in two ways. One way is with the ¢ sign. The other is with number places. Remember: cents are an example of decimal fractions (equal parts of 1). The number places for cents are the first two *decimal places*.

Example   45¢   or $. 4 5

tenths ⌐ ⌐ hundredths

You can read both expressions as "forty-five cents." When you write cents as a decimal, be sure to write the $ sign and a decimal point *before* the *dimes* place.

Exercise 2

Rewrite each amount in decimal form. The first one is done for you.

1. 11¢ = $.11

2. 25¢ = _____

3. 38¢ = _____

4. 44¢ = _____

5. 50¢ = _____

6. 75¢ = _____

7. 99¢ = _____

8. 85¢ = _____

Check your answers on page 180.

The Place-Holding Zero

Cents always need two decimal places. Some amounts will need a 0 (zero) to fill a decimal place so that the number can hold its value.

Example   1¢ = $ .0 1   10¢ = $ .1 0

         dime ⌐ ⌐ penny   dime ⌐ ⌐ penny

Exercise 3

Circle the correct decimal form for each amount.

1. 5¢ = $.05 or $.50

2. 10¢ = $.01 or $.10

3. 3¢ = $.03 or $.30

4. 80¢ = $.08 or $.80

5. 40¢ = $.04 or $.40

6. 9¢ = $.09 or $.90

Check your answers on page 180.

Total Value

You learned in Chapter 1 that the place values of a whole number add up to the number's total value. The same is true for decimal fractions.

Example $.45 is the same as 4 dimes and 5 pennies

dimes (*tenths*)   +   **pennies** (*hundredths*)

Exercise 4

What decimal place values add up to each total value?

1. $.25 = _____ *dimes* + _____ *pennies*   **5.** $.72 = _____ *dimes* + _____ *pennies*

2. $.19 = _____ *dime* + _____ *pennies*   **6.** $.07 = _____ *dimes* + _____ *pennies*

3. $.36 = _____ *dimes* + _____ *pennies*   **7.** $.88 = _____ *dimes* + _____ *pennies*

4. $.50 = _____ *dimes* + _____ *pennies*   **8.** $.91 = _____ *dimes* + _____ *penny*

Check your answers on page 180.

ON YOUR CALCULATOR

Follow these steps to enter cents:

Example   38¢ or $.38

1. Clear your calculator: C

2. Press the decimal point key: .

3. Enter the number: 3   8

4. Read the display: .38

Practice entering the decimal amounts in exercise 4 on your calculator.

Mixed Dollar Amounts

The amount at the right is a **mixed dollar** amount — dollars *and* cents.

To express mixed dollars, write a decimal point to separate dollars from cents. When reading mixed dollars, first say the dollar amount. Next say the word *and* for the decimal point. Then say the cents amount.

Example $ 1 4 . 3 4   Think: "Fourteen dollars *and* thirty-four cents."
                        ↑   ↑
whole number places   decimal places

All of the place values in a mixed dollar amount add up to the total value.
$14.34 = 1 *ten* + 4 *ones* and 3 dimes (*tenths*) + 4 pennies (*hundredths*)

Exercise 5

What amounts do the place values add up to?

1. $_____ = 3 *ones* and 5 *dimes* + 5 *pennies*

2. $_____ = 5 *tens* + 0 *ones* and 7 *dimes* + 0 *pennies*

3. $_____ = 8 *tens* + 9 *ones* and 0 *dimes* + 0 *pennies*

4. $_____ = 1 *hundred* + 5 *tens* + 3 *ones* and 6 *dimes* + 8 *pennies*

Check your answers on page 180.

ON YOUR CALCULATOR

Follow these steps to enter a mixed amount:

Example $14.34

1. Clear your calculator: [C]

2. Enter the dollar amount: [1] [4]

3. Press the decimal point key   [·]

4. Enter the cents: [3] [4]   Read the display: [14.34]

Centering on the *Ones* Place

| hundreds | tens | ones | | dimes | pennies |

The chart above shows three whole number places and the decimal places for cents. The decimal point separates whole number places from decimal places.

The decimal point *does not* hold the central position in the decimal number system. That is held by the *ones* place. The number places greater than the *ones* place are equal groups of 10 *ones*. The number places less than the *ones* place are equal parts of 1. Each group or part is always a multiple of ten.

Exercise 6

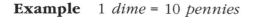

Match the values to the number places.

_____ **1.** *hundreds*   **a.** 100 equal parts of 1

_____ **2.** *tens*   **b.** equal groups of 10 *ones*

_____ **3.** *dimes*   **c.** 10 equal parts of 1

_____ **4.** *pennies*   **d.** equal groups of 100 *ones*

Check your answers on page 180.

Regrouping

You've learned that a whole number place can be regrouped and added to the next place higher or lower. That also can be done with decimal places. How?

Remember: 10 of a number place make up 1 of the next *higher* place. And 1 of a place makes up 10 of the next *lower* place.

Example   1 *dime* = 10 *pennies*

A decimal place can be regrouped as a whole number place. The same holds true for regrouping a whole number place as a decimal place.

Example   1 *one* = 10 *dimes* or $1 = 10 × $.10

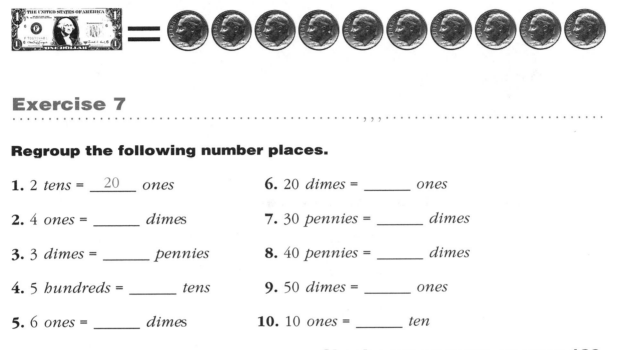

Exercise 7

Regroup the following number places.

1. 2 *tens* = ___20___ *ones*   **6.** 20 *dimes* = _____ *ones*

2. 4 *ones* = _____ *dimes*   **7.** 30 *pennies* = _____ *dimes*

3. 3 *dimes* = _____ *pennies*   **8.** 40 *pennies* = _____ *dimes*

4. 5 *hundreds* = _____ *tens*   **9.** 50 *dimes* = _____ *ones*

5. 6 *ones* = _____ *dimes*   **10.** 10 *ones* = _____ *ten*

Check your answers on page 180.

Rounding Money

You may want to round amounts to the nearest dollar, dime, or penny. When rounding to the nearest dollar, if the cents amount is less than $.50, the dollar amount stays the same. If it is $.50 or more, add $1 to the dollar amount.

Example   Round $28.15 and $28.65 to the nearest dollar.

$28.15 ≈ $28.00 $28.65 ≈ $29.00
less ↑   ↑   ↑   ↳ add $1
than $.50   stays   more
             the   than $.50
             same

Exercise 8

Round each amount to the nearest dollar.

1. $5.92 ≈ _____   **3.** $12.05 ≈ _____   **5.** $74.60 ≈ _____

2. $3.18 ≈ _____   **4.** $25.47 ≈ _____   **6.** $82.99 ≈ _____

Check your answers on page 180.

To the Nearest Dime

One way to round to the nearest dime is to look at the digit in the *pennies* place. If the digit is less than 5, round down. If the digit is 5 or more, round up.

Example Round $.13 and $.65 to the nearest dime.

$$\$. 1 \underset{\uparrow}{3} \approx \$. 1 0 \qquad \$. 6 \underset{\uparrow}{6} \approx \$. 7 0$$

less than 5,   more than 5,
round down   **round up**

Exercise 9

. . . .

Round each amount to the nearest dime.

1. $.45 ≈ _____   **4.** $.75 ≈ _____   **7.** $4.48 ≈ _____

2. $.06 ≈ _____   **5.** $1.89 ≈ _____   **8.** $5.37 ≈ _____

3. $.81 ≈ _____   **6.** $1.23 ≈ _____   **9.** $8.16 ≈ _____

Check your answers on page 180.

To the Nearest Penny

Sometimes you'll see cents worked out to three decimal places or *thousandths*. For figuring, you can round the amount to the nearest penny. You can follow the same procedure you learned for rounding to the nearest dime.

Example $1.2\underset{\uparrow}{3}6 ≈ $1.24 $1.2\underset{\uparrow}{3}4 ≈ $1.23

more than 5,   less than 5,
round up   **stays the same**

Exercise 10

. . . .

Round each amount to the nearest penny. (Remember to write the $ sign in your answer.)

1. $.188 ≈ _____   **4.** $.915 ≈ _____   **7.** $3.604 ≈ _____

2. $.453 ≈ _____   **5.** $1.473 ≈ _____   **8.** $2.508 ≈ _____

3. $.094 ≈ _____   **6.** $2.139 ≈ _____   **9.** $5.999 ≈ _____

Check your answers on page 181.

Money Checkup

How well did you understand Chapter 8?

1. Answer the questions.

   a. What is the decimal point's role (position) in a number that shows money?

   _____

   b. What decimal places do cents represent?

   _____

2. Write the names of the number places that the digits fill.

$156.19
  ↑↑ ↑ ↑↑
  a b c d e

   a. _____   **d.** _____

   b. _____   **e.** _____

   c. _____

3. What amount of money is equal to the place values?

   a. _____ = 3 _dimes_ + 5 _pennies_

   b. _____ = 4 _tens_ + 6 _ones_ and 0 _dimes_ + 8 _pennies_

   c. _____ = 6 _ones_

4. Regroup the following amounts:

   a. _____ _dimes_ = 1 _one_   **b.** _____ _pennies_ = 1 _dime_

5. Round the amounts to the nearest

   a. penny: $5.983 ≈ _____

   b. dime: $12.28 ≈ _____

   c. dollar: $18.48 ≈ _____

Check your answers on page 181.

POINTS TO REMEMBER

▶ Dollar amounts have the same place values as whole numbers. Cents have the same place values as decimal fractions.

▶ The dollar amount is to the left of the decimal point, and cents are to the right of it.

▶ Fractional dollar amounts are called _cents_, meaning _hundredths_ or 100 equal parts of one dollar. Cents always have two decimal places: they're made up of dimes (_tenths_) and pennies (_hundredths_).

CHAPTER 9 | SOLVING FOR MONEY AMOUNTS

We solve problems with money every day. Read four examples. Circle the operation you would use to get the answer and discuss why you chose it.

1. If you're paid $6.80 per hour, how much will your wages be for 40 hours of work?

   add   subtract   multiply   divide

2. How much are you paying in bills? $27.15 for garbage pickup, $22.83 for water, and $46.94 for electricity.

   add   subtract   multiply   divide

3. A microwave oven costs $119.99. If its price is reduced by $40.00, how much will it cost?

   add   subtract   multiply   divide

4. A monthly bus pass costs $30. If you use it for at least 40 bus rides, how much will each bus ride cost?

   add   subtract   multiply   divide

The answers to situations **1** and **2** need to be a *total amount*. In **1**, you *multiply* $6.80 by 40. In **2**, you *add* the three amounts.

For situation **3**, the answer has to be a *difference*. You *subtract* $40.00 from $119.99. For situation **4**, the answer must be an *equal amount* in an equal number of groups. You *divide* $30 by 40.

TALK MATH

Do these activities with a partner or group.

1. Describe some real-life situations in which you would figure money. What operation would you use in each situation?

2. With a partner, practice drilling the basic addition, subtraction, multiplication, and division facts.

Adding Money

Some addition problems may have more than one column with a sum of 10 or more.

Example $\$627 + \$986 = x$

Think: Add *ones*.

```
      1
$ 6 2 7    Regroup 13
+ 9 8 6    ones as 1 ten
$       3   and 3 ones.
```

Add *tens*.

```
    1 1
$ 6 2 7    Regroup
+ 9 8 6    11 tens as
$    1 3   1 hundred
           and 1 ten.
```

Add *hundreds*.

```
    1 1
$   6 2 7   Regroup 16 hundreds
+   9 8 6   as 1 thousand and
$ 1 6 1 3   6 hundreds.
```

Exercise 1

Solve each problem. You can check your answer on a calculator. If you get a different answer, redo the problem.

1. $\$355$
   $+179$

2. $\$577$
   $+188$

3. $\$715$
   $+485$

4. $\$106$
   $+995$

5. $\$347$
   210
   $+165$

6. $\$270$
   843
   $+579$

Check your answers on page 181.

Adding Cents

When you add cents, be sure you add the same number places. In the sum, write a $ sign and a decimal point. The decimal point should line up exactly with those in the problem.

Example $\$.39 + \$.42 + \$.18 = ?$

Think: Add *pennies*.

```
      1
$ . 3 9    Regroup
  . 4 2    19 dimes as
+ . 1 8    1 dime and
        9   9 pennies.
```

Add *dimes*.

```
      1
$ . 3 9
  . 4 2
+ . 1 8
$ . 9 9
```

Exercise 2

Solve each problem. Check each answer. (To review entering cents, reread page 78.) If you get a different answer, redo the problem.

1. $. 2 5
   + . 2 3

2. $. 4 2
   + . 2 1

3. $. 3 9
   + . 1 6

4. $. 8 3
   + . 0 7

5. $. 4 4
   . 1 3
   + . 3 0

6. $. 3 1
   . 2 5
   + . 0 7

Check your answers on page 181.

Carrying to the *Ones* Place

When the sum of the *dimes* place is 10 or more, regroup it and carry an amount to the *ones* place (one dollar). Why is that possible? Remember: the one dollar place is one place higher than the dimes place.

Example $.50 + $.75 = ?

Think: Add *pennies*.   Add *dimes*.   Add *ones*.
                                          1   1
         $. 5 0 $. 5 0   Regroup 12 $. 5 0
         + . 7 5   + . 7 5   *dimes* as 1 *one*   + . 7 5
         ─────   ─────   and 2 *dimes*.   ───────
             5   . 2 5 $1 . 2 5

Exercise 3

Solve each problem. You can check your answer on a calculator. If you get a different answer, redo the problem.

1. $. 3 5
   + . 8 2

3. $. 7 0
   + . 9 0

5. $. 6 4
   + . 5 8

7. $. 5 3
   . 6 1
   + . 3 4

2. $. 6 0
   + . 4 7

4. $. 1 8
   + . 8 5

6. $. 9 8
   + . 8 6

8. $. 4 9
   . 6 9
   + . 8 0

Check your answers on page 181.

Mixed Dollar Amounts

When setting up a problem with mixed dollar amounts, make sure you've lined up the digits in their proper number places.

Example $6.25 + $5.79 = ?

Think: Add *pennies.*

```
      1
  $ 6 . 2 5     Regroup 14
  + 5 . 7 9     pennies as
  $         4   1 dime and
                4 pennies.
```

Add *dimes.*

```
    1   1
  $ 6 . 2 5     Regroup 10 dimes
  + 5 . 7 9     as 1 one and
  $       0 4   0 dimes.
```

Add *ones.*

```
    1   1
  $ 6 . 2 5
  + 5 . 7 9
  $12. 0 4
```

Exercise 4

Solve each problem. Check your answers.

1. $1 . 2 0
   + 5 . 1 9

4. $3 . 0 8
   + 9 . 2 7

7. $1 0 . 4 7
   +   6 . 9 2

10. $1 . 5 0
      3 . 1 8
    + 4 . 0 5

2. $4 . 2 3
   + 3 . 5 6

5. $2 . 4 6
   + 5 . 8 6

8. $1 0 . 9 8
   + 2 8 . 6 9

11. $1 . 6 1
      3 . 8 5
    + 5 . 7 9

3. $5 . 4 6
   + 6 . 4 2

6. $9 . 9 9
   + 8 . 0 5

9. $2 2 . 3 5
   + 1 0 . 7 5

12. $8 . 3 7
      3 . 6 6
    + 1 2 . 5 8

Check your answers on page 181.

MATH NOTE

▶ If the answer in the *pennies* place is zero, some calculators will show only the answer to the *dimes* place. Example: .20 may be shown as .2; it's still read as 20 cents.

Subtracting Money

Before subtracting money amounts, be sure all the number places and decimal points are lined up in straight columns. The number places and the decimal point in the difference should line up with those in the problem.

When a 0 fills a number place, an amount cannot be borrowed from it. You need to go to the next higher place that's not empty to borrow 1. In the example, the first number place that an amount can be borrowed from is the *ones* place.

Example $5.00 − .82 = ?

Think: Borrow 1 *one*.

```
  4 10
$ 5 . 0 0     Regroup 1 one
−   . 8 2     as 10 dimes.
```

Borrow 1 *dime*.

```
        9
  4 10 10
$ 5 . 0 0     Regroup
−   . 8 2     1 dime as
              10 pennies.
```

Subtract.

```
        9
  4 10 10
$ 5 . 0 0
−   . 8 2
$ 4 . 1 8
```

Exercise 5

Solve each problem. You can check your answer on a calculator. If you get a different answer, redo the problem.

1. $ 1 . 0 0
    −   . 1 8

2. $ 1 . 0 0
    −   . 3 3

3. $ 5 . 0 0
    −   . 5 6

4. $ 6 . 3 6
    − 1 . 2 6

5. $ 8 . 4 9
    − 3 . 0 0

6. $ 1 2 . 6 6
    − 1 0 . 1 5

7. $ 7 . 5 3
    − 2 . 0 9

8. $ 9 . 3 8
    − 5 . 2 9

9. $ 1 4 . 7 2
    − 1 0 . 4 5

10. $ 6 . 0 0
     − 2 . 4 5

11. $ 8 . 5 2
     − 4 . 8 8

12. $ 1 5 . 0 6
     − 1 0 . 7 7

Check your answers on page 181.

Multiplying Money

Remember: in multiplication, a product of 10 or more in a number place is regrouped and carried to the next higher place. Some problems may have more than one column to be regrouped. (For a review of multiplying by *ones*, reread pages 47–52.)

Example $437 × 5 = ?$

Think: Multiply *ones*.

$$
\begin{array}{r}
3 \\
\$\ 4\ 3\ 7 \\
\times \quad\ 5 \\
\hline
\$\qquad 5
\end{array}
$$
Regroup as
3 *tens* and
5 *ones*.

Multiply *tens*.

$$
\begin{array}{r}
1\ 3 \\
\$\ 4\ 3\ 7 \\
\times \quad\ 5 \\
\hline
\$\quad 8\ 5
\end{array}
$$
Regroup as
1 *hundred*
and 8 *tens*.

$3 \times 5 = 15$ *tens*
Add: $15 + 3 = 18$ *tens*

Multiply *hundreds*.

$$
\begin{array}{r}
1\ 3 \\
\$\ 4\ 3\ 7 \\
\times \quad\ 5 \\
\hline
\$2\ 1\ 8\ 5
\end{array}
$$

$5 \times 4 = 20$ *hundreds*
Add: $20 + 1 = 21$ *hundreds*

Exercise 6

Solve each problem. You can check your answers on a calculator. If you get a different answer, redo the problem.

1. $\begin{array}{r}\$212 \\ \times\quad 8 \\ \hline\end{array}$	**3.** $\begin{array}{r}\$127 \\ \times\quad 6 \\ \hline\end{array}$	**5.** $\begin{array}{r}\$485 \\ \times\quad 3 \\ \hline\end{array}$	**7.** $\begin{array}{r}\$375 \\ \times\quad 2 \\ \hline\end{array}$
2. $\begin{array}{r}\$330 \\ \times\quad 5 \\ \hline\end{array}$	**4.** $\begin{array}{r}\$208 \\ \times\quad 7 \\ \hline\end{array}$	**6.** $\begin{array}{r}\$299 \\ \times\quad 4 \\ \hline\end{array}$	**8.** $\begin{array}{r}\$260 \\ \times\quad 9 \\ \hline\end{array}$

Check your answers on page 181.

Multiplying Cents

When you set up a multiplication problem with cents, be sure to write a $ sign and
decimal point. As you multiply the digits, remind yourself of their values. Once you
have the product of the *dimes* place, write a decimal point before it.

Example   3 × $.15 = ?

      Think: Multiply *pennies.*   Multiply *dimes.*

$$
\begin{array}{r}
1 \\
\$.15 \\
\times \quad 3 \\
\hline
5
\end{array}
\quad
\begin{array}{l}
\text{Regroup 15 } pennies \\
\text{as } \mathbf{1}\ dime \text{ and} \\
\mathbf{5}\ pennies.
\end{array}
\qquad
\begin{array}{r}
1 \\
\$.15 \\
\times \quad 3 \\
\hline
\$.45
\end{array}
\quad
\begin{array}{l}
3 \times 1 = 3\ dimes \\
\text{Add: } 3 + 1 = 4\ dimes
\end{array}
$$

Exercise 7

· ·

Solve each problem. Check your answers.

1. $.32 × 3

2. $.18 × 4

3. $.20 × 4

4. $.25 × 3

5. $.12 × 7

6. $.29 × 3

7. $.47 × 2

8. $.15 × 6

Check your answers on page 181.

Placing the Decimal Point

Remember: cents have two number places. The decimal point is written before the
dimes (or *tenths*) place. To make sure a decimal point is placed correctly in the
product, count off the number places for cents. Start counting with the digit in the
number place at the far right.

Example

$$
\begin{array}{r}
\$3.08 \\
\times \quad\quad 8 \\
\hline
\$24.64
\end{array}
\qquad\qquad
\begin{array}{r}
\$2.33 \\
\times \quad\quad 20 \\
\hline
\$46.60
\end{array}
$$

decimal point ⬑ ⬐ start
                        here
      decimal point ⬑ ⬐ start
                             here

Exercise 8

Count off the number places for cents in each product and write a decimal point. The first one is done for you.

1. 2 × $.99 = $ 1 . 9 8

2. 3 × $.75 = $ 2 2 5

3. 4 × $.33 = $ 1 3 2

4. 3 × $1.55 = $ 4 6 5

5. 2 × $4.10 = $ 8 2 0

6. 5 × $2.28 = $ 1 1 4 0

7. 10 × $4.62 = $ 4 6 2 0

8. 20 × $8.59 = $ 1 7 1 8 0

9. 12 × $.33 = $ 3 9 6

Check your answers on page 181.

Carrying to the *Ones* Place

Remember: the next higher place to the *dimes* place is the *ones* place. When the product of the *dimes* place is 10 or more, you need to regroup so an amount can be added to the *ones* place.

Example   3 × $1.45 = ?

Think: Multiply *pennies*.

```
      1
 $ 1 . 4 5     Regroup
 ×       3     15 pennies as
 ─────────     1 dime and
         5     5 pennies.
```

Multiply *dimes*.

```
   1   1
 $ 1 . 4 5     Regroup
 ×       3     13 dimes as
 ─────────     1 one and
     . 3 5     3 dimes.
```

Multiply *ones*.

```
   1   1
 $ 1 . 4 5
 ×       3
 ─────────
 $ 4 . 3 5
```

Exercise 9

Solve each problem. (Remember, there will be two decimal places for cents.) You can check your answer on a calculator. If you get a different answer, redo the problem.

1. $. 8 2
     ×   2

2. $. 7 0
     ×   4

3. $. 2 9
     ×   5

4. $. 3 6
     ×   6

5. $. 7 5
     ×   9

6. $. 9 9
     ×   3

7. $ 2 . 0 5
     ×   2

8. $ 3 . 6 3
     ×   8

9. $ 6 . 5 5
     ×   4

10. $ 8 . 1 4
      ×   5

11. $ 1 0 . 2 9
      ×   3

12. $ 1 4 . 5 7
      ×   6

Check your answers on page 181.

A Two-Place Multiplier Times Money

In Unit 1, you learned to make two multiplication problems out of one problem when the multiplier has two places. Use the same procedure when you need to multiply a dollar amount by a two-place multiplier. (To review two-place multipliers, reread page 54.)

Example Suppose you make **12 monthly installment payments** for kitchen appliances. You pay **$8.66** per month. How much do you pay altogether?

Think: Multiply by *ones*.	Multiply by *tens*.	Add the products.
1 1		1
$ 8 . 6 6	$ 8 . 6 6	$ 1 7 . 3 2
× 2	× 1 0	+ 8 6 . 6 0
$ 1 7 . 3 2	$ 8 6 . 6 0	$ 1 0 3 . 9 2

Exercise 10

Solve each equation. You can check your answers on a calculator. If you get a different answer, redo the problem.

1. 16 × $42 = _____

     $ 4 2 $ 4 2

     × 6   × 1 0

2. 15 × $67 = _____

3. 24 × $18 = _____

4. 12 × $.75 = _____

5. 18 × $.90 = _____

6. 12 × $1.05 = _____

7. 12 × $6.99 = _____

8. 24 × $12.25 = _____

9. 12 × $36.68 = _____

Check your answers on page 181.

Dividing Money

To divide money, you follow the same steps for dividing whole numbers. (To review, reread pages 67–69.) When you want to find exact change, add zeros to the dollar amount and continue dividing to show cents. Write a decimal point after the *ones* place.

Example $15 ÷ 4 = ?

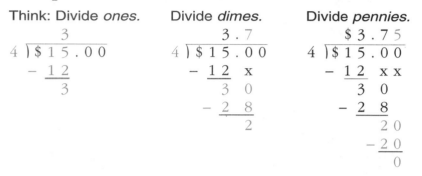

Exercise 11

Solve each problem. You can use a calculator to check the quotient. If you get a different answer, redo the problem.

1. 2)$9   **3.** 5)$18   **5.** 4)$39   **7.** 10)$72

2. 4)$17   **4.** 2)$23   **6.** 8)$52   **8.** 10)$96

Check your answers on page 181.

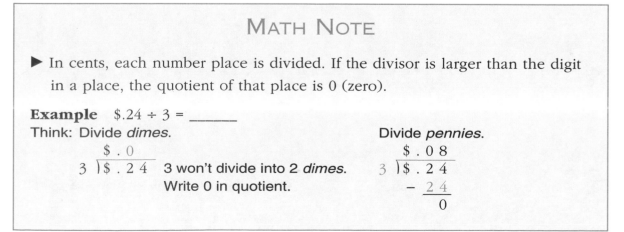

MATH NOTE

▶ In cents, each number place is divided. If the divisor is larger than the digit in a place, the quotient of that place is 0 (zero).

Example $.24 ÷ 3 = _____

Dividing Mixed Amounts

As you divide mixed amounts, remind yourself of the place value that's being divided. Make sure the number places and decimal point in the quotient line up with those in the dividend.

Example $17.94 ÷ 3 = ?

Think: Divide *ones*.

$$\begin{array}{r} 5 \\ 3\overline{)\,\$\,1\,7\,.\,9\,4} \\ -\,1\,5 \\ \hline 2 \end{array}$$

Divide: 17 *ones* ÷ 3 = 5.
Multiply: 3 x 5 = 15.
Subtract: 17 − 15 = 2 *ones*.

Divide *dimes*.

$$\begin{array}{r} 5\,.\,9 \\ 3\overline{)\,\$\,1\,7\,.\,9\,4} \\ -\,1\,5\ \ x \\ \hline 2\ 9 \\ -\,2\ 7 \\ \hline 2 \end{array}$$

29 *dimes* ÷ 3 = 9;
3 x 9 = 27;
29 − 27 = 2 *dimes*

Divide *pennies*.

$$\begin{array}{r} 5\,.\,9\,8 \\ 3\overline{)\,\$\,1\,7\,.\,9\,4} \\ -\,1\,5\ \ x\ x \\ \hline 2\ 9 \\ -\,2\ 7 \\ \hline 2\ 4 \\ -\,2\ 4 \\ \hline 0 \end{array}$$

24 *pennies* ÷ 3 = 8;
3 x 8 = 24;
24 − 24 = 0 *pennies*

Exercise 12

Solve each problem. You can use a calculator to check the quotient. If you get a different answer, redo the problem.

1. $3\overline{)\,\$\,6\,.\,1\,5}$ **4.** $6\overline{)\,\$\,1\,9\,.\,8\,0}$ **7.** $7\overline{)\,\$\,6\,9\,.\,9\,3}$ **10.** $20\overline{)\,\$\,7\,1\,.\,2\,0}$

2. $2\overline{)\,\$\,9\,.\,0\,4}$ **5.** $5\overline{)\,\$\,3\,9\,.\,4\,5}$ **8.** $12\overline{)\,\$\,2\,6\,.\,6\,4}$ **11.** $12\overline{)\,\$\,1\,2\,9\,.\,0\,0}$

3. $4\overline{)\,\$\,1\,6\,.\,6\,0}$ **6.** $8\overline{)\,\$\,5\,0\,.\,2\,4}$ **9.** $12\overline{)\,\$\,4\,0\,.\,3\,2}$ **12.** $10\overline{)\,\$\,3\,6\,1\,.\,8\,0}$

Check your answers on page 181.

Rounding a Remainder

Sometimes, you'll get a remainder after dividing the *pennies* place. You can divide one more decimal place and then round to the nearest penny. Here's what to do: Regroup the remainder as *thousandths* (the next lower decimal place). If the quotient is 5 or more, round the *pennies* place up. If the quotient is less than 5, the *pennies* place stays the same. The example shows $.59 ÷ 5.

$$
\begin{array}{r}
\$\,.118 \quad \leftarrow \text{round up to } \$.\mathbf{12} \\
5\overline{)\$\,.590} \quad \leftarrow \text{Add a zero after the } pennies \text{ place.} \\
-\underline{5}\,\text{x x} \\
09 \\
-\underline{5} \\
40 \\
-\underline{40} \\
0
\end{array}
$$

Example

Exercise 13

. . . .

Solve each problem. Round the quotient to the nearest penny. You can check your answer on a calculator. (Note: Calculator answers will need to be rounded.) If you get a different answer, redo the problem.

1. $2\overline{)\$4.53}$ **3.** $7\overline{)\$17.75}$ **5.** $5\overline{)\$29.68}$ **7.** $4\overline{)\$82.71}$

2. $3\overline{)\$5.17}$ **4.** $6\overline{)\$38.09}$ **6.** $8\overline{)\$50.90}$ **8.** $9\overline{)\$96.34}$

Check your answers on page 182.

Estimated Dollars

Remember: estimates let you know if your answers are reasonable. By using rounded numbers, you get an idea of the exact answers.

Sometimes estimates are the only answers you need. When you're shopping, for instance, you might make estimates just to be sure you have enough money.

One way of getting a dollar estimate is to round the amounts to the nearest dollar. (For a review of rounding, reread pages 81–82.) Then add, subtract, multiply, or divide the rounded figures just as you do whole numbers.

Whole milk ....$2.09 per gallon
Low-fat milk ....$2.59 per gallon

2 cartons $2.09
Colby cheese ....2 lb. for $4.99
Tastee ice cream ....16 oz. for $1.99
Sherbet, 32 oz. ....2 cartons for $5
Frozen peas ....16 oz. for $1.09
Peanut butter ....28 oz. for $3.59

$1.29
Tomato sauce, 8 oz. ....4 for $1
Pasta ....16 oz. for $.99

$2.55

Exercise 14

The problems are based on the advertisement above. Round the figures in each problem to the nearest dollar and make an estimate.

1. About how much is the total cost for each group of items?

   a. _____: pasta, four cans of tomato sauce, cheese, Tastee ice cream

   b. _____: cereal, peanut butter, low-fat milk, 2 cartons of eggs

   c. _____: bread, cheese, whole milk, frozen peas

2. About how much will the total cost be if you buy the following items?

   a. _____: 3 packages of bread

   b. _____: 2 gallons of whole milk

   c. _____: 2 packages of cheese

3. About how much is the price for one item?

   a. _____: one can of tomato sauce

   b. _____: one carton of sherbet

   c. _____: one carton of eggs

Check your answers on page 182.

Money Checkup

How well did you understand Chapter 9?

1. Make an *estimate* for each equation by rounding to the nearest dollar. Then solve the equation. (Use a calculator or solve the problems yourself.)

   a. $1.56 + $1.68 = _____   **c.** $65.73 − $29.89 = _____   **e.** $9.55 ÷ 5 = _____

   b. $3.24 + $14.79 = _____   **d.** 8 × $7.35 = _____   **f.** $80.76 ÷ 3 = _____

2. Round the amounts to the nearest penny.

   a. $14.666 = _____   **b.** $20.292 = _____   **c.** $5.333 = _____

3. Find an *estimate* for each problem.

   a. One gallon of paint costs $11.99. If you need 3 gallons, what's the approximate cost?

   Estimate: _____

   b. Suppose your total purchases cost $11.67. If you give the cashier $20.07, about how much change will you get back?

   Estimate: _____

Check your answers on page 182.

POINTS TO REMEMBER

▶ Once you've found the sum, difference, or product of the *dimes* place, write a decimal point before it.

▶ Once you've found the quotient of the *ones* place, write a decimal point after it.

▶ To make sure you've placed a decimal point correctly in a product, count off the number places for cents. The decimal point should be before the *dimes* place.

UNIT 3
DECIMALS

In this unit, you will learn basic concepts and problem solving with decimals. You'll learn how to

▶ identify decimal places and their place values

▶ recognize decimals with different names but the same value

▶ add, subtract, multiply, and divide decimals

▶ estimate answers

▶ use a calculator to get decimal answers

CHAPTER 10 | DECIMAL POWER

What are **decimals**?

Decimals, like whole numbers, are part of our number system. A decimal number can be a decimal fraction, a mixed decimal, or even a whole number.

Example $.56 $24.38 $4.00
              decimal fraction   mixed decimal   whole number

In Unit 2, you learned that we use decimals each time we deal with money. We also use decimals to measure length and distance, weight and capacity, time and temperature, and so on.

The pictures below show some measuring scales that you're probably familiar with. Study each scale. Write the type of measure that is being shown: *weight, distance, volume* (the contents of something), or *temperature*.

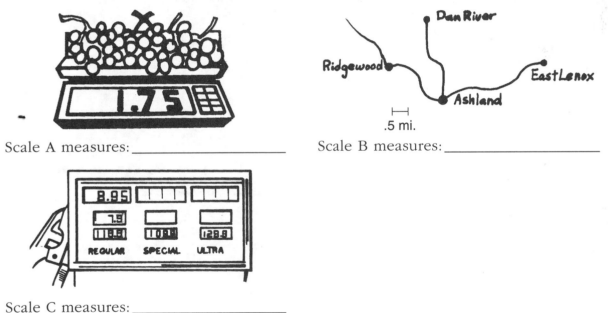

Scale A measures: _____   Scale B measures: _____

Scale C measures: _____

Did you notice that all the scales show decimals? Scale A measures weight; scale B, distance; and scale C, volume.

TALK MATH

Do these activities with a partner or group.

1. Make a list of items in your wallet or purse that have decimals on them.

2. With a partner, take turns saying a measure and giving a unit of measurement. For example, you might say, "Weight?" Your partner could answer, "Pounds."

Decimal Fractions

All fractions have a **denominator**. In Chapter 8, you learned about one denominator, *hundredths*. The denominator represents the number of equal parts that 1 whole is divided into. For decimal fractions, the denominator is always a *power of ten* — 10, 100, 1,000, etc. (The term *power* means the number of times a number is multiplied by itself.)

In decimal fractions, the number of decimal places represents the denominator.

All fractions have another part: a **numerator**. The numerator represents a certain number of equal parts out of the total represented in the denominator. In decimal fractions, the numerator is the numeral.

Example   .7 of the box is shaded.

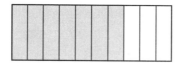

denominator: *tenths*
shaded parts: 7
unshaded parts: 3

The decimal .7 has one decimal place. So its denominator is *tenths* or 10 equal parts. Its numerator is 7, or 7 equal parts out of 10.

Exercise 1

Identify the denominator of each shape. Then identify the numerator that represents the shaded parts and the unshaded parts.

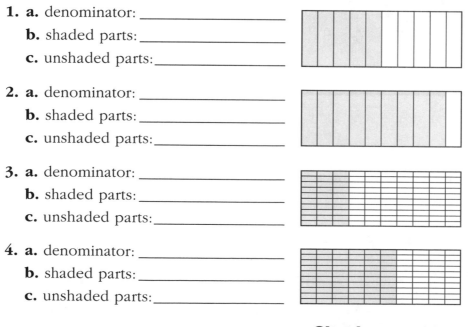

1. **a.** denominator: _____
   b. shaded parts: _____
   c. unshaded parts: _____

2. **a.** denominator: _____
   b. shaded parts: _____
   c. unshaded parts: _____

3. **a.** denominator: _____
   b. shaded parts: _____
   c. unshaded parts: _____

4. **a.** denominator: _____
   b. shaded parts: _____
   c. unshaded parts: _____

Check your answers on page 182.

Decimal Places

The chart shows the names of the first six decimal places. Read the chart from left to right. What is the first decimal place after the decimal point? What is the third decimal place? The sixth?

Tenths	Hundredths	Thousandths	Ten-Thousandths	Hundred-Thousandths	Millionths

Decimal Place Value

The shapes represent **.1** (one-*tenth*), **.01** (one-*hundredth*), and **.001** (one-*thousandth*). Which decimal fraction has the greatest value?

The *tenths* place has the greatest place value of all decimal places. The number place to the left of another is always 10 times *greater* in value. So *tenths* is 10 times greater than *hundredths*; *hundredths* is 10 times greater than *thousandths*, and so on.

The example below is a decimal fraction with three decimal places. Remember: a digit takes the value of the number place it fills. Which digit has the greatest value?

Example   . 1 2 3   ← *thousandths* place
tenths place ⤒ ⤓ *hundredths* place

.1   .01   .001

The digit 1 has the greatest value. It fills the *tenths* place.

Exercise 2

Circle the digit in color that has the greater value in each pair of numbers. Write the digit's place value. The first one is done for you.

1. .③ or .43 ___*tenths*___

2. .411 or .15 _____

3. .6 or .962 _____

4. .12 or .29 _____

5. .057 or .526 _____

6. .083 or .248 _____

7. .219 or .49 _____

8. .19 or .91 _____

Check your answers on page 182.

Expressing Decimal Fractions

Remember: the denominator is the number of equal parts that 1 has been divided into. In decimal fractions, the denominator is the total number of decimal places.

Example .589   Think: The fraction has 3 decimal places.
                     The denominator is *thousandths*.
             Read: Five-hundred-eighty-nine-*thousandths*.

Exercise 3

What is the denominator? How do you know? The first one is done for you. Read the decimal fraction aloud.

1. .15 The denominator is *hundredths*; the number has 2 decimal places.

2. .146 _____

3. .9 _____

4. .08 _____

5. .3 _____

Check your answers on page 182.

Zero as Placeholder

Some decimal fractions need a 0 to fill one or more decimal places. Why is that?

Example   *3-thousandths* is .003

Think: *Thousandths* has three decimal places; 3 fills the *thousandths* place. 0 needs to fill the *tenths* and *hundredths* places.

Exercise 4

Write the decimal fraction.

1. six-*hundredths*: _____

2. four-*thousandths*: _____

3. twelve-*thousandths*: _____

4. ninety-six-*thousandths*: _____

5. eight-*thousandths*: _____

6. nine-*hundredths*: _____

Check your answers on page 182.

On Your Calculator

Follow these steps to enter decimal fractions.

Example   .018

1. Clear your calculator: `C`

2. Press the decimal point key: `.`

3. Enter the numerator: `0` `1` `8`

4. Read the display: `.018`

Mixed Decimals

Mixed decimals are numbers that show a whole amount plus a fraction of 1.

Each of the shapes below has a value of 1 whole. The shaded parts represent the value of the mixed decimal 1.75.

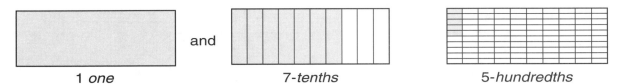

1 *one*   7-*tenths*   5-*hundredths*

The decimal point separates whole number places from decimal places. When writing a mixed decimal, the word *and* tells you where to put the decimal point.

Example   Read 1.75 as "One **and** seventy-five-*hundredths*."

Exercise 5

Write the mixed decimal.

1. Two and four-*tenths*: _____

2. Three and thirty-six-*hundredths*: _____

3. Fourteen and two-hundred-twelve-*thousandths*: _____

4. Twenty-six and ninety-nine-*hundredths*: _____

Check your answers on page 182.

Equivalent Decimals

Remember: 10 pennies is equivalent to 1 dime. Let's see how that looks as decimal fractions. The shapes below represent 1. What is the denominator of each shape?

A   **B**

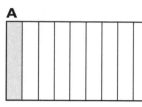

    The shaded part in shape A represents .1 (one-*tenth*). The shaded part in shape B represents .10 (ten-*hundredths*). Notice that .1 = .10.

    Decimals that represent the same value are called **equivalent decimals**.

Exercise 6

Write the decimal fractions that show equivalent amounts.

1. a. shaded parts: ___.4___ = ___.40___   **b.** unshaded parts: _____ = _____

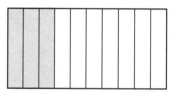

2. a. shaded parts: _____ = _____   **b.** unshaded parts: _____ = _____

3. a. shaded parts: _____ = _____   **b.** unshaded parts: _____ = _____

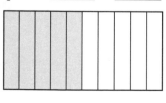

4. a. shaded parts: _____ = _____   **b.** unshaded parts: _____ = _____

Check your answers on page 182.

Renaming Decimals

You can rename a decimal so it has another denominator without changing its value. **Annex**, or attach, the number of decimal places you need for the new denominator. Write a 0 in each added decimal place. The annexed decimal places go at the end of the numerator. Remember: the digit 0 fills an empty number place and helps a number hold its value.

Example   Rename .4 (4-*tenths*) as *thousandths*.

      .4 = .4 0 0   Think: *Thousandths* has three decimal places. Annex two decimal places to the end of the numerator.

MATH NOTE

▶ Decimal places can also be annexed to a whole number. Example: Rename 15 to include 0 *hundredths*. 15 = 15.00

Exercise 7

Circle the pair of equivalent decimals in each group of numbers. The first one is done for you.

1. (.57) .057 (.570)

2. .900 .9 .09

3. .824 .82400 .08240

4. .031 3.10 3.1

5. 7.45 7.045 7.450

6. 5.00 5 .500

7. 2000 20 20.00

8. 36.30 36.3 36.03

9. 41.24 41.240 41.024

Exercise 8

Rename the numbers as equivalent amounts.

1. Rename as *hundredths*:

  a. .8 = _____
  c. 1.3 = _____
  e. 12.6 = _____
  g. 9 = _____

  b. .5 = _____
  d. 6.9 = _____
  f. 22.2 = _____
  h. 25 = _____

2. Rename as *thousandths*:

  a. .32 = _____
  c. .75 = _____
  e. 7.09 = _____
  g. 38 = _____

  b. .07 = _____
  d. 2.18 = _____
  f. 19.64 = _____
  h. 50 = _____

Check your answers on page 182.

Simplifying Decimals

Suppose you multiply numbers and get an answer that's 1.4500. But when you input the problem into the calculator, it shows an answer of 1.45. Why is that?

1.45 and 1.4500 are equivalent decimals. The calculator **simplifies** the decimal by dropping the end zeros. You can do that too. Simplified decimals are often easier to work with.

Example   Simplify 2.530

2.530 = 2.53   Think: The last decimal place is empty. Drop the 0.

Exercise 9

Simplify the decimals.

1. .360 = _____   **3.** .0400 = _____   **5.** 5.600 = _____   **7.** 28.000 = _____

2. .700 = _____   **4.** 3.00 = _____   **6.** 10.0100 = _____   **8.** 2.5050 = _____

Check your answers on page 182.

ON YOUR CALCULATOR

Follow these steps to enter mixed decimals.

Example   5.4

1. Clear your calculator: [C]

2. Enter the whole number: [5]

3. Press the decimal point key: [·]

4. Enter the numerator: [4]

5. Read the display: [5.4]

Decimal Checkup

How well did you understand Chapter 10?

1. Define each term:

    a. denominator:_____

    b. numerator:_____

    c. equivalent decimals: _____

2. Match the decimal place to its value.

    _____ *tenths*   **a.** 1,000 equal parts of 1

    _____ *hundredths*   **b.** 10 equal parts of 1

    _____ *thousandths* **c.** 100 equal parts of 1

3. Circle the number place that has the lesser value.

    a. *tenths* or *hundredths*

    b. *hundredths* or *thousandths*

    c. *thousandths* or *ten-thousandths*

    d. *tenths* or *ones*

4. Rename the following numbers as *thousandths*.

    a. .25 = _____   **b.** 15.9 = _____   **c.** 3 = _____

5. Simplify the decimals.

    a. 4.890 = _____   **b.** .80 = _____   **c.** .070 = _____   **d.** 15.000 = _____

Check your answers on page 183.

POINTS TO REMEMBER

▶ The first three decimal places in order from greatest to least value are *tenths*, *hundredths*, and *thousandths*.

▶ The decimal point separates the whole number from the decimal fraction.

▶ Equivalent decimals have the same value.

▶ Rename a decimal fraction as an equivalent decimal by annexing one or more decimal places to the end of the numerator.

▶ Simplify decimals by dropping end zeros.

CHAPTER 11 | SOLVING DECIMAL PROBLEMS

In real life, we often need to solve problems with decimals. Read the four examples. Circle the operation that you would use to get the answer. Then discuss why you would use that operation.

a. Normal body temperature is 98.6 degrees. Suppose your temperature rises to 101 degrees. How much above normal is your temperature?

add   subtract   multiply   divide

b. Suppose you buy 3 packages of ground round. They weigh 1.01 pounds, .97 pounds, and .87 pounds. How many pounds are there altogether?

add   subtract   multiply   divide

c. A car travels 150.8 miles on 4.6 gallons of gasoline. How many miles does the car get per gallon?

add   subtract   multiply   divide

d. A yard of silk costs $12.99. If you buy .75 yard, how much do you pay?

add   subtract   multiply   divide

The answers to situations **b** and **d** must be a *total amount*. In **b**, you would add the three amounts. In **d**, you would multiply the amounts.

You would subtract the amounts in situation **a**. That's because the answer has to be a *difference*. You would divide the amounts in situation **c**. The answer to that question should be an *equal amount*.

In this chapter, you will use your basic operation skills to figure decimals. And you'll continue to learn how to solve and check problems.

Adding Decimals

Often an addition problem will have decimals with different denominators. To help you keep track of adding the correct decimal places, rename the numbers so that they have the same denominator.

As you add the digits, remind yourself of their place value. When it's needed, regroup and carry the amount to the next higher place. Remember to write a decimal point between the *tenths* and *ones* places in the sum.

Example   Suppose it rained over three days. On Sunday, it rained **.5 inches**; on Monday, **1 inch**; and on Tuesday, **.75**. What's the total rainfall?

Think: Add *hundredths*.

$$\begin{array}{r} .50 \\ 1.00 \\ +.75 \\ \hline 5 \end{array}$$
Rename decimals as hundredths.

Add: $0 + 0 + 5 = 5$

Add *tenths*.

$$\begin{array}{r} ^1 \\ .50 \\ 1.00 \\ +.75 \\ \hline .25 \end{array}$$
Regroup 12 *tenths* as 1 *one* and 2 *tenths*.

Add: $5 + 0 + 7 = 12$

Add *ones*.

$$\begin{array}{r} ^1 \\ .50 \\ 1.00 \\ +.75 \\ \hline 2.25 \end{array}$$

Add: $1 + 1 = 2$

Exercise 1

Rewrite the equations as problems. Rename the decimals so they have the same denominator. Then solve the problems. Check your answers.

1. $.5 + .17 =$ _____

$$\begin{array}{r} .50 \\ +.17 \\ \hline \end{array}$$

3. $3.694 + 8.1 =$ _____

5. $7.6 + 9 + 12.09 =$ _____

2. $.099 + .45 =$ _____

4. $4 + .759 =$ _____

6. $.653 + .8 + 1.4 =$ _____

Check your answers on page 183.

Subtracting Decimals

When subtracting decimals, be sure all number places and decimal points are lined up. If decimals have different denominators, first rename them with the same denominator and then subtract.

    As you subtract digits, remind yourself of their place values. Remember: if the top digit is smaller than the digit being subtracted, you need to regroup an amount from the next higher place. (Sometimes you may need to regroup from two or more places over.)

Example   Let's say you estimate you'll type a report in **2.25** hours. You finish it in **1.5 hours**. How much sooner was the report done?

Think: Subtract *hundredths.*

$$\begin{array}{r} 2.2^5 \\ -1.5^0 \\ \hline 5 \end{array}$$
Rename decimals as *hundredths.*

Subtract *tenths.*

$$\begin{array}{r} {}^{1}\,{}^{12} \\ \cancel{2}.\cancel{2}5 \\ -1.5\,0 \\ \hline .7\,5 \end{array}$$
Borrow 1 *one*; regroup as 10 *tenths*. Add: 10 + 2 = 12.

Subtract *ones.*

$$\begin{array}{r} {}^{1}\,{}^{12} \\ \cancel{2}.\cancel{2}5 \\ -1.5\,0 \\ \hline .7\,5 \end{array}$$
Ones place can show a 0 or not.

Exercise 2

Rewrite the equations as problems. Rename the decimals so they have the same denominator. Then solve the problems. Check your answers.

1. .84 − .3 = _____

4. 3.855 − .73 = _____

7. 10.3 − 2.898 = _____

2. .5 − .209 = _____

5. 4 − 1.16 = _____

8. 21.57 − 16.4 = _____

3. .77 − .614 = _____

6. 9.9 − 7.35 = _____

9. 20 − 15.553 = _____

Check your answers on page 183.

Multiplying Decimals

The procedure for multiplying decimals is the same as for multiplying money. With decimals, though, the values you're multiplying can become confusing.

Here's one way to handle decimal problems. Ignore the decimal points when you multiply. Imagine you are multiplying whole numbers.

Once you have the product, you can determine where the decimal point should go. Count the total decimal places in the problem. Then count off the same number of decimal places in the answer. Start counting with the digit in the far right number place.

Example $.3 \times 2.3 = ?$

Think: Multiply.　　　Place decimal point.

```
   2 . 3          2 . 3     1 decimal place
 ×   . 3        ×   . 3   + 1 decimal place
   6 9            . 6 9     2 decimal places
start count here ↗
```

Exercise 3

. . . .

Write a decimal point in each product. Explain why. The first one is done for you.

1.
```
   2 . 5
 ×     2
   5 . 0
```
The problem has 1 decimal place.

4.
```
   1 1 . 2 2
 ×       . 4
   4 4 8 8
```
_____

2.
```
   . 4 5
 ×     3
   1 3 5
```
_____

5.
```
   9 . 2 0
 ×   . 2 5
   2 3 0 0 0
```
_____

3.
```
   . 7 5 9
 ×       8
   6 0 7 2
```
_____

6.
```
   6 . 1
 ×   . 3 2
   1 9 5 2
```
_____

Check your answers on page 183.

The Multiplier

When the multiplier in a decimal problem is a whole number, the answer you get is greater than the original amount.

Exercise 4

Solve each equation. Use a calculator or do your work on another sheet of paper.

1. $2 \times 3.15 =$ _____   **4.** $4 \times 4.74 =$ _____   **7.** $4 \times 8.6 =$ _____

2. $3 \times 6.1 =$ _____   **5.** $2 \times 7.025 =$ _____   **8.** $10 \times 9.64 =$ _____

3. $5 \times .65 =$ _____   **6.** $3 \times 10.9 =$ _____   **9.** $40 \times .29 =$ _____

Check your answers on page 183.

A Decimal Fraction Multiplier

When the multiplier is a decimal, the answer you need is a *part* of the original amount. In the example, the answer would be a part (.5) of $3.20.

Example What is .5 of $3.20?

Think: Multiply.   Place decimal point.

$$
\begin{array}{r}
1 \\
3.20 \\
\times \quad .5 \\
\hline
1.600
\end{array}
\qquad
\begin{array}{r}
3.20 \\
\times \quad .5 \\
\hline
1.600 \\
\uparrow
\end{array}
\qquad
\begin{array}{l}
\text{2 decimal places} \\
\text{+ 1 decimal place} \\
\hline
\text{3 decimal places}
\end{array}
$$

$1.600 = **$1.60**   start count here

Exercise 5

Solve each equation. Use a calculator or do your work on another sheet of paper. If you get a different answer, redo the problem.

1. $.3 \times 15 =$ _____   **4.** $3 \times 1.07 =$ _____   **7.** $.10 \times 14 =$ _____

2. $.2 \times 28 =$ _____   **5.** $.5 \times .7 =$ _____   **8.** $.20 \times 9 =$ _____

3. $.2 \times 2.6 =$ _____   **6.** $.3 \times .43 =$ _____   **9.** $.10 \times 5.03 =$ _____

Check your answers on page 183.

Two-Place Multiplier

When the multiplier is a two-place number, remember you can solve the problem by creating two problems out of one. (For a review, reread page 54.) With decimals, do one extra step: rewrite the problem with whole numbers. Do this before making the two problems.

Once you find the final product, determine where the decimal point should go by counting the number of decimal places in the original problem.

Example   A suitcase costs **$69**. The store puts it on sale for **.25** of the regular price. What is the sale price?

Think: Rewrite problem.   Multiply.   Add products.   Place decimal point.

```
                              4     1              1
    6 9      6 9             6 9    6 9           3 4 5                  6 9
  × . 2 5  ×  2 5          ×   5  ×  2 0        + 1 3 8 0            ×  . 2 5    2 decimal places
                            3 4 5  1 3 8 0         1 7 2 5           1 7 . 2 5   start count here
  original  as whole       multiply multiply                            ↑
  problem   numbers        ones     tens
```

Exercise 6

Solve each equation. Check your answers. If you get a different answer, redo the problem. The first one has been started for you.

1. .33 × 7.6 = _____

```
    3 3      3 3      3 3        1 9 8
  × 7 6    ×   6    × 7 0    + 2 3 1 0
    1 9 8    2 3 1 0         2 5 0 8
```

6. .18 × 12.72 = _____

2. 3.1 × 5.8 = _____

7. 1.4 × 5.21 = _____

3. .75 × 2.9 = _____

8. .45 × 72 = _____

4. 2.2 × 34.9 = _____

9. .15 × 8.4 = _____

5. 12 × 20.39 = _____

10. 1.7 × 5.03 = _____

Check your answers on page 183.

Dividing Decimals

As you've learned, when you have a remainder in division, you can divide further. Annex one or more zeros to the end of the dividend.

Some problems will always have a remainder. In those problems, you can round to a chosen number place. This includes remainders that are repeating decimals. For example, $2 \div 3 = .666 \ldots$

Example   Suppose you're planning a **3.8-mile** hike. You want **5 rest stops** the same distance apart. How many miles will be between stops?

Think: Divide *tenths*.

$$\begin{array}{r} .7 \\ 5\overline{)3.8} \\ -3\,5 \\ \hline 3 \end{array}$$

Place the decimal point in the quotient over the decimal point in the problem.

Divide: $38 \div 5 = 7$ *tenths*
Multiply: $5 \times 7 = 35$ *tenths*
Subtract: $38 - 35 = 3$

Divide *hundredths*.

$$\begin{array}{r} .7\,6 \text{ mile} \\ 5\overline{)3.8\,0} \\ -3\,5\,x \\ \hline 3\,0 \\ -3\,0 \\ \hline 0 \end{array}$$

Annex 1 number place.

Divide: $30 \div 5 = 6$
Multiply: $5 \times 6 = 30$
Subtract: $30 - 30 = 0$

Exercise 7

Solve the problems. Round any repeating decimals to the *thousandths* place. Check your answers. If you get a different answer, redo the problem.

1. $2\overline{)4.4}$ **4.** $2\overline{)12.19}$ **7.** $5\overline{)7.3}$ **10.** $12\overline{)36.8}$

2. $5\overline{)3.5}$ **5.** $4\overline{)49.4}$ **8.** $10\overline{)8.56}$ **11.** $15\overline{)75.3}$

3. $4\overline{).15}$ **6.** $6\overline{)20.3}$ **9.** $12\overline{)9.06}$ **12.** $20\overline{)42.35}$

Check your answers on page 183.

Dividing by a Decimal

To do a division math problem with a decimal divisor, first you need to rename the divisor as a whole number. To change decimals to whole numbers, multiply by 10 (or a power of ten). When you rename the divisor, you also must rename the dividend. Multiply the dividend by the same amount.

Example $4.8 \div .8 = ?$

Rename divisor.	Rename dividend.	Write division problem.
.8 One decimal	4 . 8	$8\overline{)48}$
×10 place, so	× 10	
8 . 0 multiply by 10.	4 8 . 0 ← drop end 0	
↑		
drop end zero		

Example $4.8 \div .08 = ?$

Rename divisor.	Rename dividend.	Write division problem.
.0 8 Two decimal	4 . 8	$8\overline{)480}$
×100 places, so	× 100	
8 . 0 0 multiply by 100.	4 8 0 . 0 ← drop end zero	
↑		
drop end zeros		

Remember: what power of ten you multiply by depends on the number of decimal places in the divisor.

Exercise 8

. . . .

Rename each division problem so the divisor is a whole number. Show how you changed the problem. Simplify the decimals before dividing. The first one is done for you. Do not solve the problems.

1. $.15\overline{)3.72}$ is renamed as $\underline{15\overline{)372}}$

  $100 \times .15 = 15$

  $100 \times 3.72 = 372$

2. $.3\overline{)4.25}$ is renamed as _____

_____

_____

3. $1.4\overline{)17.2}$ is renamed as _____

_____

_____

4. $.9\overline{)8.1}$ is renamed as _____

_____

_____

Check your answers on page 183.

Solving the Problem

Once you've renamed the decimal divisor as a whole number (and made the same changes to the dividend), you're ready to solve the problem. (Note: when using a calculator to solve division problems, you do not need to rename either the divisor or the dividend.)

Example   Suppose you do a temporary job. You receive **$45.27** for **4.5 hours** of work. How much did you make per hour?

Think: Rename problem.   Divide *tens*.   Divide *ones, tenths*.   Divide *hundredths*.

$$4.5 \overline{)\, \$\,45.27} \rightarrow 45 \overline{)\, \$452.7}$$

$$
\begin{array}{r}
1 \\
45\overline{)\,\$\,4\,5\,2.7} \\
-\underline{4\,5} \\
0
\end{array}
\qquad
\begin{array}{r}
1\,0.0 \\
45\overline{)\,\$\,4\,5\,2.7} \\
-\underline{4\,5}\,\text{x x} \\
0\,2\,7
\end{array}
\qquad
\begin{array}{r}
\$\,1\,0.0\,6 \\
45\overline{)\,\$\,4\,5\,2.7\,0} \\
-\underline{4\,5}\,\text{x x x} \\
0\,2\,7\,0 \\
-\underline{2\,7\,0} \\
0
\end{array}
$$

Exercise 9

Divide to three decimal places and round the quotients to the nearest hundredth. You can check your answers on a calculator.

1. 2.4 ÷ .2 = _____

5. 18 ÷ .5 = _____

9. 31.25 ÷ 2.5 = _____

2. 9.1 ÷ .8 = _____

6. 45 ÷ .09 = _____

10. 12.9 ÷ 4.3 = _____

3. 1.23 ÷ .5 = _____

7. 3.55 ÷ .07 = _____

11. 7.92 ÷ .33 = _____

4. .963 ÷ .3 = _____

8. .36 ÷ .18 = _____

12. 65 ÷ .15 = _____

Check your answers on page 184.

Estimated Decimals

One way to estimate problems with decimals is to round mixed decimals to the nearest whole number. Then do the math. If the amounts are all decimal fractions, round the figures to the nearest *tenth*.

Example: 3.5 + 14.6 = ?

$$\begin{array}{r} 3.5 \approx 4 \\ + 14.6 \approx + 15 \\ \hline 19 \end{array} \leftarrow \text{estimate}$$

Exercise 10

Read each situation. Write the solution. Then estimate the answer. (Use the problem-solving strategy to help you choose the correct math operation.)

1. A recipe calls for a 3.5-pound chicken. If you buy a 3.128-pound chicken, how much less will you have than the recipe calls for?

2. A recipe calls for 8 ounces of tomato paste. The store sells only 3.125-ounce cans. How many ounces do two cans total?

3. You have a 24.3-fluid ounce can of juice. The label says there are 6 servings in a can. How many ounces is each serving?

4. A cheese shop is having a sale. You buy 1.79 pounds of Monterey Jack and 2 pounds of cheddar. How many pounds of cheese do you buy in all?

5. The trip odometer in your car says 314.9 miles. If you've been driving for 5 hours, about how far have you driven every hour?

Check your answers on page 184.

Decimal Checkup

How well did you understand Chapter 11?

1. In problem A, why are 14 and 24.57 written with extra decimal places?

_____

_____

_____

A. $14 + 24.57 + 9.845 = ?$

```
  1 1  1
  1 4 . 0 0 0
  2 4 . 5 7 0
+   9 . 8 4 5
─────────────
  4 8 . 4 1 5
```

2. In problem B, how is the amount borrowed for the *tenths* place?

_____

_____

_____

B. $30.7 - 18.95 = ?$

```
    9 16
  2 10 6 10
  3̶0̶ . 7̶ 0̶
- 1 8 . 9 5
───────────
  1 1 . 7 5
```

3. In problem C, is the decimal point in the correct place in the product? Explain why.

_____

_____

_____

C. $3.5 \times 12.5 = 43.75$

4. In problem D, show how the divisor and dividend are renamed?

_____

_____

_____

D. $9.6 \div .5 = ?$

```
        1 9.2
    5 ) 9 6.0
      - 5 x
      ─────
        4 6 x
      - 4 5
      ─────
          1 0
        - 1 0
        ─────
            0
```

Check your answers on page 184.

POINTS TO REMEMBER

▶ When decimals have different denominators in addition and subtraction problems, rename the decimal fractions so they have the same denominator.

▶ When the divisor is a decimal, rename it as a whole number. Multiply both the divisor and the dividend by the same power of ten.

UNIT 4
COMMON
FRACTIONS

In this unit, you will learn basic concepts and problem solving with common fractions. You'll learn how to

▶ describe the terms of a common fraction

▶ express amounts with common fractions and mixed numbers

▶ rename common fractions as equivalent (equal) fractions

▶ add, subtract, multiply, and divide common fractions

▶ estimate answers

CHAPTER 12 FRACTION POWER

When 1 whole is divided into equal parts, we sometimes use **fractions** to talk about those equal parts. In Units 2 and 3, you learned about one form of fraction — *decimal fractions*. They name equal parts of 1 that have been divided into *tenths*, *hundredths*, and other groups that are powers of ten.

Another form of fraction is called **common fractions** (also known just as fractions). With common fractions, 1 whole can be divided into any number of equal parts. Probably the most common fraction is $\frac{1}{2}$ (one-half).

 The shape represents 1 whole divided into 2 equal parts. Each equal part is $\frac{1}{2}$ of 1.

A whole amount or 1 whole that is divided into equal parts may be **1 unit**.

half an apple

half a dollar

half a cup

half an inch

A whole amount also may be **1 group** of things.

half a dozen eggs

half a six-pack of soda pop

In this unit, you'll learn to express amounts as common fractions. You'll also learn to compare fractions and to round them to whole numbers.

TALK MATH

Do these activities with a partner or group.

1. Describe several situations in which you would use common fractions.

2. With a partner, take turns giving an example of half of 1 whole and saying what the whole is. For example, you might say, "Half an hour." Your partner should say, "The whole is one hour."

Equal Parts of 1

Remember, 1 whole can be divided into any number of equal parts. You can prove that by folding a sheet of paper.

Get a sheet of paper. The paper represents 1 whole. Fold the paper in half, matching its ends. (See Figure A.) Unfold the sheet. How many equal parts do you see?

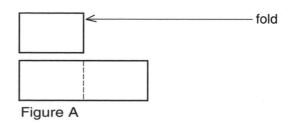

Figure A

1 whole is divided into two equal parts or two **halves** ($\frac{2}{2}$). Each equal part is one-half ($\frac{1}{2}$).

Fourths

Refold the paper; then fold it in half again. (See Figure B.) Unfold the sheet. How many equal parts are there?

1 whole is divided into 4 equal parts or four **fourths** ($\frac{4}{4}$). Each equal part is one-fourth ($\frac{1}{4}$).

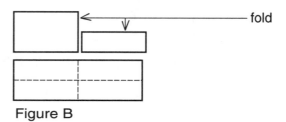

Figure B

Eighths

Refold the paper. Then fold it one more time. (See Figure C.) Unfold the sheet. How many equal parts is 1 whole divided into?

You should see 8 equal parts or eight **eighths** ($\frac{8}{8}$). Each equal part is one-eighth ($\frac{1}{8}$).

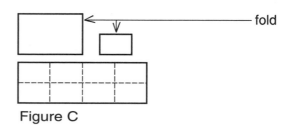

Figure C

Exercise 1

Match each shape to its number of equal parts.

a. fourths (4 equal parts)   **b.** fifths (5 equal parts)   **c.** eighths (8 equal parts)

_____ 1. 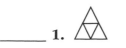   _____ 2.   _____ 3. ⊛

Check your answers on page 184.

Naming Equal Parts

Every fraction has a numerator and a denominator. That is, it has two **terms**. The numerator represents the number of equal parts being discussed. The denominator represents the total number of equal parts that 1 whole is divided into.

Example   Four-*tenths* of the shape is shaded. Six-*tenths* is unshaded.

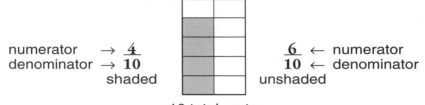

numerator   → 4   6 ← numerator
denominator → 10   10 ← denominator
        shaded   unshaded

10 total parts

Common fractions are written with the numerator over the denominator. A line bar separates the two numbers. To read a common fraction, first say the numerator and then say the denominator: "four-*tenths*, six-*tenths*."

How would you express 1 whole as a common fraction? In the example, $\frac{10}{10}$ equals a whole, or 1. Other fractions that equal 1 are $\frac{1}{1}$, $\frac{2}{2}$, $\frac{3}{3}$, etc.

Exercise 2

Each shape represents 1 whole. Write the common fraction that names the whole, the shaded portion, and the unshaded portion.

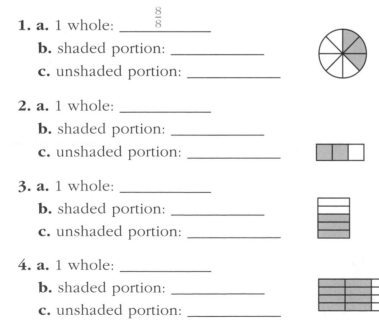

1. a. 1 whole: ____$\frac{8}{8}$____

    b. shaded portion: _____

    c. unshaded portion: _____

2. a. 1 whole: _____

    b. shaded portion: _____

    c. unshaded portion: _____

3. a. 1 whole: _____

    b. shaded portion: _____

    c. unshaded portion: _____

4. a. 1 whole: _____

    b. shaded portion: _____

    c. unshaded portion: _____

Check your answers on page 184.

The Greater Value

The fewer equal parts that 1 whole is divided into, the greater the value of the denominator.

Example   Which fraction has the greater value: $\frac{1}{2}$ or $\frac{1}{10}$?

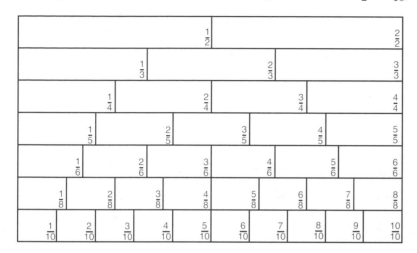

The **fraction table** shows halves, thirds, fourths, fifths, sixths, eighths, and tenths.

As you can see in the table, $\frac{1}{2}$ is greater than $\frac{1}{10}$. Why? Because halves are larger equal parts than tenths.

Exercise 3

Circle the fraction that has the greater value. Then explain why with words or show it with pictures. The first one is done for you.

1. $\frac{1}{3}$ or $\frac{1}{4}$ $\frac{1}{3}$; thirds are larger equal parts than fourths

2. $\frac{1}{2}$ or $\frac{1}{4}$ _____

3. $\frac{1}{8}$ or $\frac{1}{3}$ _____

4. $\frac{1}{5}$ or $\frac{1}{2}$ _____

5. $\frac{1}{16}$ or $\frac{1}{8}$ _____

Check your answers on page 184.

Comparing the Same Numerators and Denominators

When common fractions have different denominators but the same numerators, you can easily figure out which fraction is greater.

Example   Which is greater: $\frac{2}{3}$ or $\frac{2}{4}$?

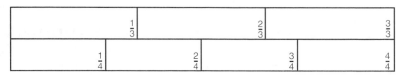

When comparing fractions with the same numerator but a different denominator, the fraction with the smaller denominator is greater. $\frac{2}{3}$ is greater than $\frac{2}{4}$.

You can easily compare fractions that have the same denominators. The fraction with the larger numerator has the greater value.

Example   Which is greater: $\frac{3}{8}$ or $\frac{5}{8}$?

Both denominators are the same, so the fraction with the larger numerator is greater. $\frac{5}{8}$ is greater than $\frac{3}{8}$.

Exercise 4

Read each situation. Circle the fraction that shows the larger amount.

1. A bread recipe calls for $\frac{1}{4}$ cup wheat bran and $\frac{3}{4}$ cup oat flour.

2. In a political poll, $\frac{2}{3}$ of the respondents are Democrats. In a different poll, $\frac{2}{5}$ of the respondents are Democrats.

3. Typing makes up $\frac{3}{5}$ of the total job. Editing and proofing make up $\frac{1}{5}$ of the job.

4. At a hardware store, you can buy $\frac{3}{8}$-inch nails or $\frac{3}{4}$-inch nails.

5. Aaron's commute takes $\frac{1}{2}$ hour by bus or $\frac{1}{3}$ hour by subway.

6. One piece of wood measures $\frac{7}{12}$ foot. A second piece measures $\frac{3}{12}$ foot. The third piece measures $\frac{9}{12}$ foot.

Check your answers on page 184.

Improper Fractions

When the numerator of a fraction is equal to or larger than the denominator, the fraction is called an **improper fraction**.

When the numerator is *larger than* the denominator, the fraction may represent a whole amount.

Example $\frac{12}{6}$ = 2 tins of muffins (1 tin = $\frac{6}{6}$)

$\frac{12}{6}$ ← number of equal parts discussed
← total equal parts that 1 is divided into

Exercise 5

Write the improper fraction that names each whole amount.

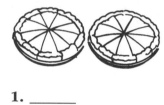  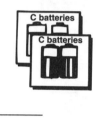

1. _____   2. _____   3. _____

Check your answers on page 184.

Naming Mixed Amounts

When the numerator is *larger than* the denominator, the fraction represents a mixed amount, a whole plus a fraction.

Example $\frac{9}{6}$ = 1 ($\frac{6}{6}$) tin of muffins + $\frac{3}{6}$ of 1 tin

$\frac{9}{6}$ ← number of equal parts discussed
← total equal parts that 1 is divided into

Exercise 6

Write the improper fraction that names each portion.

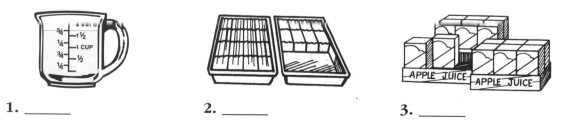

1. _____   2. _____   3. _____

Check your answers on page 184.

Mixed Numbers

On the last page, you saw how a mixed amount is expressed as an improper fraction. Another way to express a mixed amount is with a **mixed number**. You write a whole number and a common fraction to represent the amount.

Example $\frac{9}{2} = 4\frac{1}{2}$ Think: "Four *and* one-half."

Exercise 7

Write the mixed number for each mark pointed to on the scale.

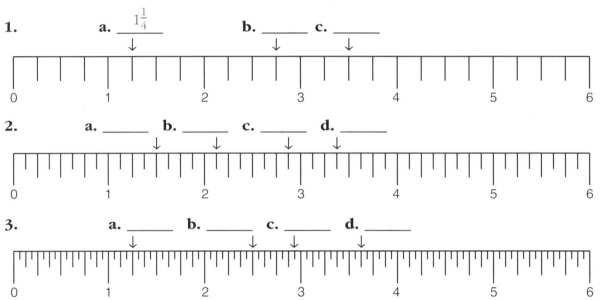

1.   a. $1\frac{1}{4}$ b. _____   c. _____

2.   a. _____   b. _____   c. _____   d. _____

3.   a. _____   b. _____   c. _____   d. _____

Exercise 8

Circle the mixed amount; then write it as a mixed number.

$1\frac{1}{3}$ **1.** The drive between San Juan and Whitney takes one and one-third hours.

_____ **2.** A box of detergent weighing four and three-fourths pounds costs $6.99.

_____ **3.** In six months, a child grew three and five-sixteenths inches.

_____ **4.** For shelves, you need three pieces of wood that are each three and seven-tenths feet long.

_____ **5.** The freeway sign shows six and one-half miles to the downtown exit.

Check your answers on page 184.

More or Less than $\frac{1}{2}$

When you estimate answers, it helps to know if a fraction is more or less than $\frac{1}{2}$.

▶ A fraction is less than $\frac{1}{2}$ if the numerator is less than half of the denominator.

Example $\frac{3}{8}$ is less than $\frac{1}{2}$ Think: Half of 8 (denominator) is 4.

The numerator is less than 4.

▶ A fraction is more than $\frac{1}{2}$ if the numerator is more than half of the denominator.

Example $\frac{5}{8}$ is more than $\frac{1}{2}$ Think: Half of 8 (denominator) is 4.

The numerator is greater than 4.

Rounding to the Nearest Whole Number

In estimation, you may want to round mixed numbers to the nearest whole number. Here's one way to round mixed numbers:

▶ If the fraction is less than $\frac{1}{2}$, round the whole number down.

▶ If the fraction is $\frac{1}{2}$ or more than $\frac{1}{2}$, round the whole number up.

Example Round $2\frac{1}{3}$ and $3\frac{3}{5}$ to the nearest whole numbers.

$2\frac{1}{3} \approx 2$ $2\frac{3}{5} \approx 3$

Think: $\frac{1}{3}$ is less than $\frac{1}{2}$;   Think: $\frac{3}{5}$ is greater than $\frac{1}{2}$;
   round whole number down.   round whole number up.

Exercise 9

Round each mixed number to the nearest whole number.

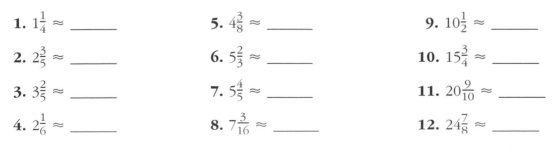

1. $1\frac{1}{4} \approx$ _____

2. $2\frac{3}{5} \approx$ _____

3. $3\frac{2}{5} \approx$ _____

4. $2\frac{1}{6} \approx$ _____

5. $4\frac{3}{8} \approx$ _____

6. $5\frac{2}{3} \approx$ _____

7. $5\frac{4}{5} \approx$ _____

8. $7\frac{3}{16} \approx$ _____

9. $10\frac{1}{2} \approx$ _____

10. $15\frac{3}{4} \approx$ _____

11. $20\frac{9}{10} \approx$ _____

12. $24\frac{7}{8} \approx$ _____

Check your answers on page 184.

Fraction Checkup

How well did you understand Chapter 12?

1. Check each sentence below that correctly describes the amount shown. (There may be more than one answer.)

   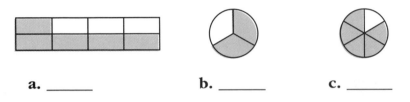

   ☐ **a.** $\frac{3}{4}$ is less than 1

   ☐ **b.** $\frac{3}{4}$ out of $\frac{4}{4}$

   ☐ **c.** The denominator 3 shows how many equal parts 1 whole is divided into.

2. The shapes below each represent 1 whole. What fraction names the shaded portion of each object?

   a. _____   **b.** _____   **c.** _____

3. Write the common fraction or mixed number that names a mark on the scale.

   ```
   0                      1                      2
   |  |  |  |  |  |  |  |  |  |  |  |  |  |  |  |
        ↑           ↑              ↑
   ```

   a. _____   **b.** _____   **c.** _____

4. Circle the fraction that has the lesser value in each pair of fractions.

   a. $\frac{1}{4}$ $\frac{3}{4}$ **b.** $\frac{2}{5}$ $\frac{3}{5}$ **c.** $\frac{1}{2}$ $\frac{1}{3}$ **d.** $\frac{3}{8}$ $\frac{5}{8}$

5. Read each sentence. Circle the mixed amount; then round it to the nearest whole number. Write the rounded number on the blank before the sentence.

   _____ **a.** You work two days of overtime for a total of six and one-half hours.

   _____ **b.** Five large apples weigh two and three-fourths pounds.

Check your answers on page 184.

PONTS TO REMEMBER

▶ A whole amount may be 1 unit or 1 group of things.

▶ Common fractions are expressed by writing two terms: the numerator over the denominator.

▶ Fractions that have the numerator equal to or greater than the denominator are called *improper fractions*. They show whole or mixed amounts.

▶ Mixed numbers represent a whole amount plus a fractional amount.

CHAPTER 13 | RENAMING FRACTIONS

In Unit 3, you learned to *rename* or change decimals without changing the value of the number.

Common fractions can also be renamed without changing their value. When two or more fractions show the same amount, they are *equivalent*. Let's see how. Get a sheet of paper.

Fold the sheet to get halves. Unfold the sheet. Label the left portion $\frac{1}{2}$. (See Figure A.)

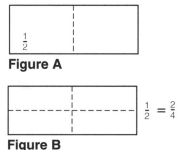

Figure A

Now refold the sheet in half, then fold it in half again. Unfold the sheet. Each folded portion is $\frac{1}{4}$. How many fourths equal $\frac{1}{2}$?

The fractions $\frac{2}{4}$ and $\frac{1}{2}$ show the same amount. Write the amount that is equivalent to your label: $\frac{1}{2} = \frac{2}{4}$. (See Figure B.)

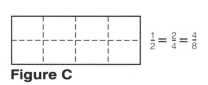

$\frac{1}{2} = \frac{2}{4}$

Figure B

Now fold your paper to make eighths. Unfold the sheet. Each folded portion is $\frac{1}{8}$. How many eighths equal $\frac{1}{2}$ and $\frac{2}{4}$? Write the amount that is equivalent to your label. (See Figure C.)

$\frac{1}{2} = \frac{2}{4} = \frac{4}{8}$

Figure C

The fraction $\frac{4}{8}$ shows the same amount as $\frac{1}{2}$ and $\frac{2}{4}$. Thus, $\frac{4}{8}$, $\frac{1}{2}$, and $\frac{2}{4}$ are all equivalent fractions.

In this chapter, you'll learn to rename common fractions, mixed numbers, and improper fractions. And you'll learn to change fractions to decimals as well as decimals to fractions.

TALK MATH

Do these activities with a partner or group.

1. Using your folded sheet of paper, prove that $\frac{2}{8} = \frac{1}{4}$ and that $\frac{6}{8} = \frac{3}{4}$.

2. Refold the paper and fold it in half again. How many equal parts is 1 divided into? What fraction is equivalent to $\frac{1}{2}$? $\frac{1}{4}$? $\frac{3}{4}$?

Equivalent Fractions

The shapes at the right each represent 1 whole. The shaded portion in each shape is the same amount. Write the equivalent fractions.

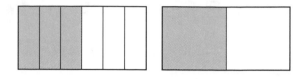

The fractions $\frac{3}{6}$ and $\frac{1}{2}$ show the same amount. They are **equivalent fractions**.

Exercise 1

All the objects below represent 1 whole. Write the equivalent fractions that represent the shaded portions.

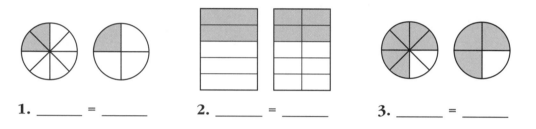

1. _____ = _____

2. _____ = _____

3. _____ = _____

Check your answers on page 184.

Getting an Equivalent Fraction

You can get an equivalent fraction by raising the terms of the fraction to new terms. Here's how: multiply both the denominator and numerator by the same number.

Example $\frac{1}{2} = \frac{2}{4}$

Think: $\frac{1 \times 2}{2 \times 2} = \frac{2}{4}$

$\frac{1}{2} = \frac{3}{6}$

Think: $\frac{1 \times 3}{2 \times 3} = \frac{3}{6}$

$\frac{1}{2} = \frac{4}{8}$

Think: $\frac{1 \times 4}{2 \times 4} = \frac{4}{8}$

Exercise 2

Find the equivalent fractions. Remember to multiply the numerator *and* the denominator by the same amount.

1. Multiply by 2:

   a. $\frac{1}{4}$ **b.** $\frac{2}{3}$ **c.** $\frac{3}{8}$

2. Multiply by 3:

   a. $\frac{1}{4}$ **b.** $\frac{2}{3}$ **c.** $\frac{3}{8}$

3. Multiply by 4:

   a. $\frac{1}{4}$ **b.** $\frac{2}{3}$ **c.** $\frac{3}{8}$

4. Multiply by 5:

   a. $\frac{1}{4}$ **b.** $\frac{2}{3}$ **c.** $\frac{3}{8}$

Check your answers on page 185.

Renaming with a Given Denominator

Sometimes you'll need to rename a fraction as an equivalent fraction with a *given* denominator. Follow the steps below using the given denominator.

Example $\frac{2}{3} = \frac{?}{12}$ 12 is the given denominator.

Think: Divide denominators.	Multiply numerator by quotient.	Complete the equivalent fractions.
$3\overline{)12}$ with quotient 4	$\begin{array}{r} 2 \\ \times\, 4 \\ \hline 8 \end{array}$	$\frac{2}{3} = \frac{8}{12}$

Exercise 3

Rename the fractions with the given denominator. Show all your work.

1. $\frac{1}{2} = \dfrac{}{6}$

2. $\frac{1}{3} = \dfrac{}{6}$

3. $\frac{3}{4} = \dfrac{}{8}$

4. $\frac{2}{5} = \dfrac{}{10}$

5. $\frac{2}{5} = \dfrac{}{15}$

Check your answers on page 185.

Finding a Common Denominator

To add or subtract fractions, the denominators must be alike. If they're different, find a **common denominator** (having the same denominator) in one of two ways:

▶ If one denominator can be divided evenly by the other denominator(s), then that is the common denominator.

Example $\frac{1}{8}$, $\frac{1}{2}$ Think: The denominator 2 can divide evenly into the other denominator 8. Common denominator: 8.

▶ Multiply the denominators to get the common denominator.

Example $\frac{1}{2}$, $\frac{1}{3}$ Think: 3 x 2 = 6
Common denominator: 6.

Exercise 4

Find a common denominator for each pair of fractions.

1. $\frac{1}{2}, \frac{1}{4}$: _____

2. $\frac{1}{3}, \frac{1}{6}$: _____

3. $\frac{1}{4}, \frac{2}{3}$: _____

4. $\frac{2}{5}, \frac{1}{3}$: _____

5. $\frac{3}{4}, \frac{2}{5}$: _____

6. $\frac{3}{10}, \frac{4}{5}$: _____

7. $\frac{1}{2}, \frac{5}{6}$: _____

8. $\frac{2}{3}, \frac{1}{12}$: _____

Check your answers on page 185.

Reducing Fractions

Each shape below represents 1 whole. The same amount is shaded in each shape. In the first shape, the shaded portion can be written as $\frac{4}{8}$. How can the shaded portion in the other shape be written? Write the fraction below the shape.

$$\frac{4}{8} \qquad = \qquad \underline{\hspace{1.5cm}}$$

The fractions $\frac{4}{8}$ and $\frac{2}{4}$ are equivalent. They both show the same amount. Often you can **reduce** a fraction to an equivalent fraction with lower terms. To reduce a fraction, find a number that divides evenly into both numerator and denominator.

Example $\quad \frac{2}{4} = \frac{1}{2} \quad \frac{2 \div 2}{4 \div 2} = \frac{1}{2} \quad$ Think: 2 can divide evenly into 2 and 4; $2 \div 2 = 1$, $4 \div 2 = 2$; reduced equivalent fraction is $\frac{1}{2}$.

Exercise 5

Write a number that can divide evenly into both numerator and denominator. The first one is done for you. Do not solve.

1. $\frac{3 \div 3}{6 \div 3}$ **2.** $\frac{5 \div}{10 \div}$ **3.** $\frac{6 \div}{16 \div}$ **4.** $\frac{8 \div}{10 \div}$ **5.** $\frac{16 \div}{24 \div}$

Check your answers on page 185.

To Its Lowest Terms

"Reduce fractions to their lowest terms" means to divide the numerator and denominator until they can no longer be divided evenly. This may be done in one step or many steps.

Example   Reduce $\frac{12}{24}$ to its lowest terms.

$$\frac{12 \div 2}{24 \div 2} = \frac{6}{12} \qquad \frac{6 \div 3}{12 \div 3} = \frac{2}{4} \qquad \frac{2 \div 2}{4 \div 2} = \frac{1}{2} \quad \text{OR} \quad \frac{12 \div 12}{24 \div 12} = \frac{1}{2}$$

Exercise 6

Reduce the fractions to their lowest terms. Show all your work.

1. $\frac{8}{16}$ **3.** $\frac{16}{24}$ **5.** $\frac{18}{24}$ **7.** $\frac{15}{45}$

2. $\frac{6}{18}$ **4.** $\frac{12}{20}$ **6.** $\frac{24}{36}$ **8.** $\frac{18}{48}$

Check your answers on page 185.

Common Fractions = Decimal Fractions

A common fraction and a decimal fraction can name the same amount. To change a decimal fraction to a common fraction, write the numerator over the denominator. (Remember: the numerator in decimal fractions is the number and the denominator is the total of decimal places.) If needed, reduce to lowest terms.

Example   .5 = ?

Rename decimal.	Reduce to lowest terms.	Equivalents
$.5 = \dfrac{5}{10}$ One decimal place equals *tenths*.	$\dfrac{5 \div 5}{10 \div 5} = \dfrac{1}{2}$	$.5 = \dfrac{1}{2}$

Exercise 7

Rename the decimal fractions as common fractions. Reduce the fractions to their lowest terms. Show all your work.

1. .8   **2.** .25   **3.** .6   **4.** .84   **5.** .75   **6.** .125

Check your answers on page 185.

Renaming as Decimal Fractions

To rename a common fraction as a decimal fraction, divide the numerator by the denominator. Since the numerator is smaller than the denominator, you'll need to annex one or more zeros to fill decimal places.

Example $\dfrac{1}{2}$ = ?

$$\begin{array}{r} .5 \\ 2\overline{)\,1.0} \\ -\underline{1\,0} \\ 0 \end{array}$$

Think: Divide the numerator by the denominator: 1 ÷ 2. Annex 0 to the *tenths* place. Regroup 1 *one* as 10 *tenths*.

Exercise 8

Rename the common fractions as decimal fractions. Round repeating decimals to the nearest *hundredth*.

1. $\dfrac{1}{4}$ = _____   **3.** $\dfrac{1}{8}$ = _____   **5.** $\dfrac{3}{8}$ = _____   **7.** $\dfrac{3}{5}$ = _____

2. $\dfrac{3}{4}$ = _____   **4.** $\dfrac{1}{3}$ = _____   **6.** $\dfrac{5}{6}$ = _____   **8.** $\dfrac{2}{3}$ = _____

Check your answers on page 185.

RENAMING FRACTIONS | 133

Mixed Numbers = Improper Fractions

A mixed amount can be expressed as either a mixed number or an improper fraction. For example, $2\frac{1}{4}$ is the same as $\frac{9}{4}$.

Exercise 9

Write the improper fraction and mixed number that express each shaded portion.

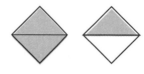

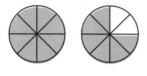

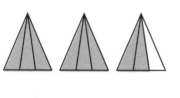

1. _____ = _____   2. _____ = _____   3. _____ = _____

Check your answers on page 185.

Renaming as an Improper Fraction

Here's how to rename a mixed number as an improper fraction:

Example $2\frac{1}{4} = ?$

Keep the same denominator.	Multiply the denominator by the whole number.	Add the product to the numerator.
	$4 \times 2 = 8$	$8 + 1 = 9$
$2\frac{1}{4} = \frac{}{4}$	$2\frac{1}{4} = \frac{}{4}$	$2\frac{1}{4} = \frac{9}{4}$

Exercise 10

Rename the mixed number as an improper fraction. Show your work.

1. $4\frac{1}{4} =$ _____   3. $1\frac{2}{3} =$ _____   5. $3\frac{1}{2} =$ _____   7. $2\frac{1}{3} =$ _____

2. $2\frac{1}{2} =$ _____   4. $1\frac{3}{4} =$ _____   6. $4\frac{2}{5} =$ _____   8. $3\frac{7}{8} =$ _____

Check your answers on page 185.

Renaming Improper Fractions

Usually, if an answer you solve for is an improper fraction, you'll need to **simplify** it. That is, change it to a whole number or mixed number.

Example $\frac{9}{2}$ = ?

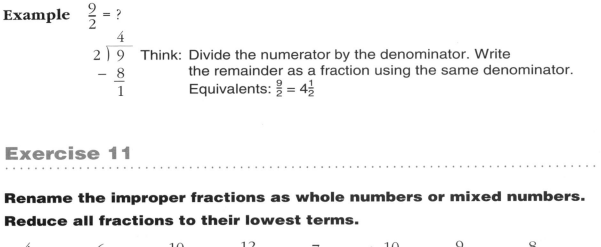

Think: Divide the numerator by the denominator. Write the remainder as a fraction using the same denominator. Equivalents: $\frac{9}{2} = 4\frac{1}{2}$

Exercise 11

Rename the improper fractions as whole numbers or mixed numbers. Reduce all fractions to their lowest terms.

1. $\frac{4}{2}$ 2. $\frac{6}{3}$ 3. $\frac{10}{10}$ 4. $\frac{12}{4}$ 5. $\frac{7}{2}$ 6. $\frac{10}{3}$ 7. $\frac{9}{4}$ 8. $\frac{8}{5}$

Check your answers on page 185.

Rename as Mixed Decimals

To rename improper fractions as mixed decimals, continue dividing the remainder. How would you do that? (To review dividing for a decimal amount, see page 114.)

Example $\frac{9}{2}$ = ?

Think: Divide the numerator by the denominator. Annex 0 in the *tenths* place. Rename 1 *one* as 10 *tenths*. Equivalents: $\frac{9}{2} = 4.5$

Exercise 12

Use a calculator to rename the fractions (divide the numerator by the denominator). Round repeating decimals to the nearest *hundredth*.

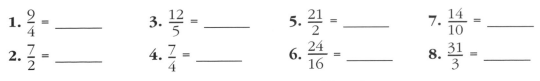

1. $\frac{9}{4}$ = _____   3. $\frac{12}{5}$ = _____   5. $\frac{21}{2}$ = _____   7. $\frac{14}{10}$ = _____

2. $\frac{7}{2}$ = _____   4. $\frac{7}{4}$ = _____   6. $\frac{24}{16}$ = _____   8. $\frac{31}{3}$ = _____

Check your answers on page 185.

Fraction Checkup

How well did you understand Chapter 13?

1. Write the missing numerators in each set of equivalent fractions.

a. $\frac{1}{2} = \frac{}{4} = \frac{}{6} = \frac{}{8} = \frac{}{10}$

b. $\frac{1}{3} = \frac{}{6} = \frac{}{9} = \frac{}{12} = \frac{}{15}$

2. Name a common denominator for each pair of fractions.

a. $\frac{3}{5}, \frac{1}{3}$: _____   b. $\frac{1}{2}, \frac{7}{8}$: _____

3. Reduce the fractions to their lowest terms. Show all your work.

a. $\frac{8}{16}$ b. $\frac{3}{12}$ c. $\frac{8}{10}$

4. Rename the mixed numbers as improper fractions. Show all your work.

a. $2\frac{1}{2}$ b. $3\frac{3}{4}$ c. $4\frac{1}{5}$

5. Rename the improper fractions as mixed numbers. Show all your work.

a. $\frac{9}{2}$ b. $\frac{8}{5}$ c. $\frac{14}{3}$

6. Use a calculator to help you rename the fractions as decimals.

a. $\frac{1}{4} =$ _____   b. $\frac{2}{5} =$ _____   c. $1\frac{3}{8} =$ _____   d. $5\frac{1}{2} =$ _____

Check your answers on page 185.

POINTS TO REMEMBER

▶ Equivalent fractions have the same value.

▶ To raise the terms of a common fraction, multiply the numerator and denominator by the same amount.

▶ To reduce the terms, divide the numerator and denominator by the same amount.

▶ To rename a common fraction as a decimal fraction, divide the numerator by the denominator.

CHAPTER 14 | **FRACTIONS AND MIXED NUMBERS**

In daily life, we often use common fractions and mixed numbers. For instance, we use fractions and mixed numbers when we go grocery shopping. We might read them on packages or on the store's scale when we weigh fruits or vegetables. What's another example of using fractions and mixed numbers when we shop for food?

Many of us use common fractions to discuss time. We may talk about doing something in $\frac{1}{4}$ of an hour or $1\frac{1}{2}$ hours. We even use fractions to talk about longer times such as $\frac{1}{3}$ of a year or $4\frac{3}{4}$ years.

When stores have a sale, they often use common fractions to show how much prices are discounted. Think of the many sale signs that advertise $\frac{1}{2}$ off, $\frac{1}{3}$ off, or $\frac{1}{4}$ off.

In this chapter, you'll learn the basic problem-solving concepts and skills to add, subtract, multiply, and divide common fractions and mixed numbers.

TALK MATH

Do these activities with a partner or group.

1. Describe a situation in which you need to use common fractions or mixed numbers.

2. With a partner, take turns describing where you have seen fractions being used.

Adding Common Fractions

You would add numbers such as fractions because you need to find a total amount. Remember: you can add only amounts that are alike. With fractions, the denominators also must be alike.

Example A cookie recipe calls for $\frac{1}{4}$ **cup** whole wheat flour, $\frac{1}{4}$ **cup** oat flour, and $\frac{1}{4}$ **cup** white flour. How many cups of flour are combined?

Solving the Problem

A solution for common fractions can be expressed as an equation or a problem. In a problem, you may want to write the numbers in a row (horizontally) for easier adding. That's because only the numerators are added. The sum will have the *same denominator* as the fractions being added.

Example $\frac{1}{4} + \frac{1}{4} + \frac{1}{4} = \frac{3}{4}$ Think: Add numerators: 1 + 1 + 1 = 3.
Don't add the denominators.

Exercise 1

Is the numerator correct in each sum? If the sum is incorrect, write the correct answer. Then write the denominator for each sum.

1. $\frac{1}{5} + \frac{3}{5} = \frac{4}{\rule{1cm}{0.4pt}}$

2. $\frac{7}{12} + \frac{4}{12} = \frac{12}{\rule{1cm}{0.4pt}}$

3. $\frac{5}{16} + \frac{8}{16} = \frac{14}{\rule{1cm}{0.4pt}}$

4. $\frac{3}{10} + \frac{6}{10} = \frac{9}{\rule{1cm}{0.4pt}}$

5. $\frac{1}{8} + \frac{3}{8} + \frac{2}{8} = \frac{9}{\rule{1cm}{0.4pt}}$

6. $\frac{1}{3} + \frac{2}{3} + \frac{2}{3} = \frac{5}{\rule{1cm}{0.4pt}}$

Exercise 2

Solve the equations. Simplify (reduce) the sums.

1. $\frac{1}{4} + \frac{2}{4} =$

2. $\frac{3}{6} + \frac{2}{6} =$

3. $\frac{4}{8} + \frac{2}{8} =$

4. $\frac{2}{5} + \frac{1}{5} =$

5. $\frac{1}{2} + \frac{1}{2} =$

6. $\frac{2}{3} + \frac{2}{3} =$

7. $\frac{3}{16} + \frac{5}{16} =$

8. $\frac{3}{4} + \frac{3}{4} =$

9. $\frac{1}{2} + \frac{1}{2} + \frac{1}{2} =$

10. $\frac{1}{8} + \frac{3}{8} + \frac{2}{8} =$

11. $\frac{4}{10} + \frac{5}{10} + \frac{1}{10} =$

12. $\frac{4}{8} + \frac{6}{8} + \frac{2}{8} =$

Check your answers on page 185.

Adding Unlike Denominators

Remember: you can add fractions only when they have the same denominator. If the denominators are **unlike**, or different, you'll need to find a common denominator. Then rename some or all the fractions with the common denominator. (To review renaming to a common denominator, reread page 131.)

Example   Suppose you have $\frac{1}{2}$ **quart** of engine oil left in a can. In another can you have $\frac{1}{4}$ **quart** of engine oil. How much oil do you have in all?

Think: Find a common denominator.

$$\frac{1}{2} \quad \frac{1}{4} \qquad \begin{array}{l} 4 \div 2 = 2 \\ \text{So, 4 is the} \\ \text{common denominator.} \end{array}$$

Rename fraction(s).

$$\frac{1}{2} = \frac{2}{4} \qquad \begin{array}{l} \text{Multiply numerator by 2.} \\ \text{Multiply denominator by 2.} \end{array}$$

Add.

$$\frac{2}{4} + \frac{1}{4} = \frac{3}{4}$$

Exercise 3

Find the common denominator in each equation. Then rewrite the equation using the common denominator and add. Simplify the sums.

1. $\frac{1}{2} + \frac{3}{4} = ?$

2. $\frac{2}{3} + \frac{1}{2} = ?$

3. $\frac{4}{5} + \frac{1}{2} = ?$

4. $\frac{3}{8} + \frac{1}{4} = ?$

5. $\frac{2}{5} + \frac{7}{10} = ?$

6. $\frac{3}{4} + \frac{1}{6} = ?$

7. $\frac{3}{4} + \frac{1}{3} = ?$

8. $\frac{1}{2} + \frac{5}{8} = ?$

9. $\frac{5}{16} + \frac{3}{4} = ?$

10. $\frac{1}{2} + \frac{1}{4} + \frac{3}{8} = ?$

11. $\frac{1}{3} + \frac{3}{4} + \frac{5}{12} = ?$

12. $\frac{1}{4} + \frac{1}{3} + \frac{1}{2} = ?$

Check your answers on page 185.

Adding Mixed Numbers

When adding mixed numbers, rename the fractions as needed. First add the fractions, then add the whole numbers. Reduce fraction sum to lowest terms.

Example $2\frac{1}{2} + 1\frac{3}{4} = ?$

Think: Rename fractions with a common denominator.

$\frac{3}{4}$

$\frac{1}{2} = \frac{2}{4}$

Since 2 can divide into 4, 4 is the common denominator. Rename $\frac{1}{2}$ as fourths.

Add fractions and whole numbers.

$2\frac{2}{4}$
$+ 1\frac{3}{4}$
$\overline{3\frac{5}{4}}$

Rename improper fraction as a mixed number and add.

$3\frac{5}{4} = 3 + 1\frac{1}{4} = 4\frac{1}{4}$

Exercise 4

Solve the equations. Simplify all answers.

1. $2\frac{1}{4} + 3\frac{1}{4} = ?$

2. $4\frac{1}{8} + 2\frac{3}{8} = ?$

3. $1\frac{3}{10} + 3\frac{5}{10} = ?$

4. $1\frac{1}{2} + 3\frac{1}{4} = ?$

5. $2\frac{1}{3} + 2\frac{1}{2} = ?$

6. $4\frac{2}{4} + 3\frac{1}{4} = ?$

7. $4\frac{3}{8} + 2\frac{5}{8} = ?$

8. $5\frac{1}{2} + 2\frac{1}{2} = ?$

9. $3\frac{3}{4} + 4\frac{1}{4} = ?$

10. $1\frac{1}{2} + 4\frac{5}{8} = ?$

11. $3\frac{3}{4} + 4\frac{1}{2} = ?$

12. $2\frac{2}{3} + 5\frac{1}{2} = ?$

Check your answers on page 186.

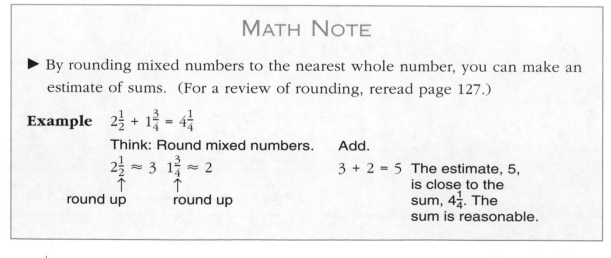

MATH NOTE

▶ By rounding mixed numbers to the nearest whole number, you can make an estimate of sums. (For a review of rounding, reread page 127.)

Example $2\frac{1}{2} + 1\frac{3}{4} = 4\frac{1}{4}$

Think: Round mixed numbers.   Add.

$2\frac{1}{2} \approx 3 \quad 1\frac{3}{4} \approx 2$
　　↑　　　　↑
round up　　round up

$3 + 2 = 5$ The estimate, 5, is close to the sum, $4\frac{1}{4}$. The sum is reasonable.

Subtracting Common Fractions

With fractions, subtract only the numerators. You can subtract fractions if they have a common denominator. Reduce all fractions to their lowest terms.

Example $\frac{3}{4} - \frac{1}{4} = ?$

Subtract numerators.

$\frac{3}{4} - \frac{1}{4} = \frac{2}{4}$ Denominators are not subtracted.

Reduce to lowest terms.

$\frac{2}{4} = \frac{1}{2}$ Divide both terms (2 and 4) by 2.

Exercise 5

Solve the problems. Reduce all fractions to their lowest terms.

1. $\frac{2}{3} - \frac{1}{3} =$

2. $\frac{6}{8} - \frac{3}{8} =$

3. $\frac{10}{12} - \frac{3}{12} =$

4. $\frac{14}{16} - \frac{8}{16} =$

5. $\frac{9}{10} - \frac{5}{10} =$

Check your answers on page 186.

Unlike Denominators

When a subtraction problem has unlike denominators, you need to find a common denominator. Then rename one or both fractions with that common denominator.

Example $\frac{1}{2} - \frac{1}{3} = ?$

Find a common denominator.

$\frac{1}{2} \quad \frac{1}{3} \quad 2 \times 3 = 6$

common ↑ denominator

Rename fractions.

$\frac{1 \times 3}{2 \times 3} = \frac{3}{6}$

$\frac{1 \times 2}{3 \times 2} = \frac{2}{6}$

Subtract fractions.

$\frac{3}{6} - \frac{2}{6} = \frac{1}{6}$

Exercise 6

Solve the equations. Reduce all fractions to their lowest terms.

1. $\frac{3}{4} - \frac{1}{2} = ?$

2. $\frac{1}{2} - \frac{1}{3} = ?$

3. $\frac{4}{5} - \frac{3}{10} = ?$

4. $\frac{2}{3} - \frac{1}{2} = ?$

5. $\frac{5}{8} - \frac{1}{4} = ?$

6. $\frac{2}{3} - \frac{1}{5} = ?$

7. $\frac{3}{4} - \frac{3}{16} = ?$

8. $\frac{1}{2} - \frac{5}{12} = ?$

Check your answers on page 186.

Subtracting from a Whole Number

Sometimes, you'll need to subtract a fraction from a whole number. One way to solve such a problem is to regroup 1 *one*. You rename the amount as an improper fraction with the same denominator.

Example   Suppose you plan to walk **2 miles**. So far, you've walked $\frac{3}{4}$ **mile**. How many more miles will you need to walk?

Think: Borrow 1 *one*.   Rename with common denominator.   Subtract.

$\overset{1}{\cancel{2}}\frac{1}{1}$ Regroup 1 as $\frac{1}{1}$. $1\frac{4}{4}$ Rename $\frac{1}{1}$ as fourths. $1\frac{4}{4}$

$-\frac{3}{4}$ $-\frac{3}{4}$ $-\frac{3}{4}$

$1\frac{1}{4}$

Exercise 7

Rename 1 as an improper fraction with the given denominator.

1. 1 as thirds   **3.** 1 as fifths   **5.** 1 as eighths   **7.** 1 as fourths

_____   _____   _____   _____

2. 1 as halves   **4.** 1 as tenths   **6.** 1 as sixteenths   **8.** 1 as sixths

_____   _____   _____   _____

Exercise 8

Solve the equations. Reduce the fractions to their lowest terms.

1. $1 - \frac{2}{3} = ?$ **4.** $3 - \frac{4}{5} = ?$ **7.** $9 - \frac{1}{2} = ?$

2. $1 - \frac{3}{4} = ?$ **5.** $5 - \frac{3}{4} = ?$ **8.** $7 - \frac{2}{3} = ?$

3. $2 - \frac{3}{8} = ?$ **6.** $3 - \frac{9}{10} = ?$ **9.** $8 - \frac{1}{2} = ?$

Check your answers on page 186.

Subtracting Mixed Numbers

When subtracting mixed numbers, subtract the common fractions first. You may need to rename unlike denominators with a common denominator or regroup 1 *one* before subtracting. Reduce the difference of the fractions to its lowest terms. Finally, subtract the whole numbers.

Example $4\frac{1}{4} - 2\frac{3}{8} = ?$

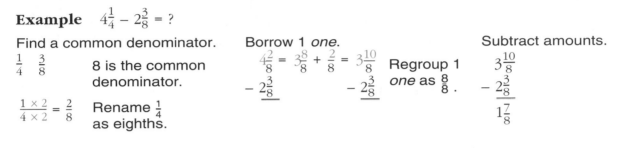

Find a common denominator.

$\frac{1}{4}$ $\frac{3}{8}$ 8 is the common denominator.

$\frac{1 \times 2}{4 \times 2} = \frac{2}{8}$ Rename $\frac{1}{4}$ as eighths.

Borrow 1 *one*.

$4\frac{2}{8} = 3\frac{8}{8} + \frac{2}{8} = 3\frac{10}{8}$ Regroup 1 *one* as $\frac{8}{8}$.

$-2\frac{3}{8}$ $-2\frac{3}{8}$

Subtract amounts.

$3\frac{10}{8}$
$-2\frac{3}{8}$
$1\frac{7}{8}$

Exercise 9

Solve the equations. Reduce the fractions to their lowest terms.

1. $2\frac{2}{3} - 1\frac{1}{3} = ?$ **4.** $5\frac{1}{2} - 1\frac{1}{4} = ?$ **7.** $8\frac{1}{4} - 3\frac{3}{4} = ?$ **10.** $3 - 2\frac{1}{2} = ?$

2. $4\frac{3}{4} - 2\frac{1}{4} = ?$ **5.** $4\frac{1}{2} - 1\frac{3}{8} = ?$ **8.** $7\frac{1}{5} - 4\frac{3}{5} = ?$ **11.** $7\frac{5}{8} - 5\frac{1}{2} = ?$

3. $1\frac{4}{5} - 1\frac{3}{5} = ?$ **6.** $5\frac{1}{2} - 2\frac{1}{3} = ?$ **9.** $6\frac{1}{2} - 3\frac{2}{3} = ?$ **12.** $9\frac{1}{4} - 8\frac{1}{2} = ?$

Check your answers on page 186.

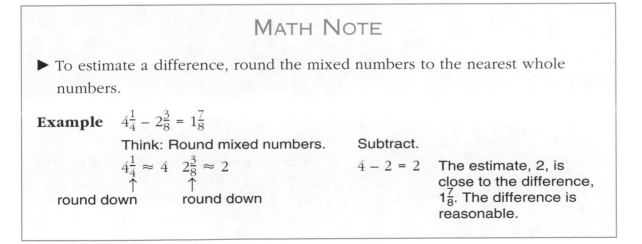

MATH NOTE

▶ To estimate a difference, round the mixed numbers to the nearest whole numbers.

Example $4\frac{1}{4} - 2\frac{3}{8} = 1\frac{7}{8}$

Think: Round mixed numbers.   Subtract.

$4\frac{1}{4} \approx 4$ $2\frac{3}{8} \approx 2$ $4 - 2 = 2$ The estimate, 2, is close to the difference, $1\frac{7}{8}$. The difference is reasonable.

round down   round down

Multiplying Common Fractions

When you're multiplying to find a fraction of an amount, the answer will be a smaller amount.

Example A recipe calls for $\frac{1}{2}$ cup milk. If you divide the recipe in $\frac{1}{2}$, how much milk will you then need?

The example is asking "What is $\frac{1}{2}$ of $\frac{1}{2}$ cup?" The word *of* is a clue that you must multiply the amounts.

To solve, set up a multiplication problem. Then, multiply the numerators and then the denominators. Reduce fractions to their lowest terms.

$\frac{1}{2}$ of $\frac{1}{2}$ cup $\frac{1}{2} \times \frac{1}{2} = ?$

Solution: Multiply numerators.   Multiply denominators.
$\frac{1}{2} \times \frac{1}{2} = \frac{1}{-}$ $\frac{1}{2} \times \frac{1}{2} = \frac{1}{4}$

Exercise 10

Set up the solution for each problem. Then solve the problem. Reduce fractions to lowest terms.

1. $\frac{1}{2}$ of $\frac{3}{4}$ yard of fabric
Solution:

4. $\frac{1}{2}$ of $\frac{1}{2}$ of a dollar
Solution:

2. $\frac{1}{4}$ of $\frac{1}{2}$ gallon of milk
Solution:

5. $\frac{1}{4}$ of $\frac{1}{4}$ pound of cheese
Solution:

3. $\frac{1}{3}$ of $\frac{9}{10}$ mile
Solution:

6. $\frac{1}{3}$ of $\frac{3}{4}$ hour
Solution:

Check your answers on page 186.

A Fraction of a Whole Amount

When you need to find a fraction of a whole amount, rename the whole number as a fraction using 1 as the denominator.

Example You estimate it will take **6 hours** to pack everything in the kitchen. Hired movers estimate $\frac{3}{4}$ **of that time**. How long do the movers think it will take?

Rename whole number.

$\frac{3}{4} \times \frac{6}{1} = ?$

Multiply.

$\frac{3}{4} \times \frac{6}{1} = \frac{18}{4}$

Simplify answer.

$\frac{18 \div 2}{4 \div 2} = \frac{9}{2} = 4\frac{1}{2}$ **hours**

When an answer is an improper fraction, simplify the amount. Either rename it as a mixed number or reduce to lowest terms.

Exercise 11

Solve the equations. Simplify all answers. Show your work.

1. $\frac{1}{2} \times 8 = ?$

5. $\frac{1}{4} \times 5 = ?$

9. $\frac{3}{4} \times 7 = ?$

2. $\frac{1}{2} \times 12 = ?$

6. $\frac{1}{3} \times 9 = ?$

10. $\frac{2}{3} \times 15 = ?$

3. $\frac{1}{2} \times 9 = ?$

7. $\frac{1}{5} \times 12 = ?$

11. $\frac{3}{8} \times 12 = ?$

4. $\frac{1}{2} \times 15 = ?$

8. $\frac{1}{10} \times 8 = ?$

12. $\frac{2}{5} \times 10 = ?$

Check your answers on page 186.

A Fraction of a Mixed Amount

Here's one way to find a fraction of a mixed number. First, rename the mixed number as an improper fraction. (For a review of that skill, reread page 134.) Then, multiply the numerators and denominators.

Example   Suppose you lose $8\frac{1}{2}$ **pounds**. If you lost $\frac{1}{4}$ **of that amount** in one week, how much did you lose?

Rename mixed number.   Multiply.   Simplify answer.

$\frac{1}{4} \times 8\frac{1}{2} = ?$ $\frac{1}{4} \times \frac{17}{2} =$ $\frac{1}{4} \times \frac{17}{2} = \frac{17}{8}$ $\frac{17}{8} = 2\frac{1}{8}$

Exercise 12

Solve the equations. Simplify all answers. Show your work.

1. $\frac{1}{2} \times 1\frac{1}{2} = ?$ **4.** $\frac{3}{4} \times 5\frac{1}{2} = ?$ **7.** $\frac{1}{3} \times 5\frac{1}{4} = ?$

2. $\frac{1}{3} \times 3\frac{1}{2} = ?$ **5.** $\frac{2}{3} \times 6\frac{1}{3} = ?$ **8.** $\frac{1}{2} \times 6\frac{3}{4} = ?$

3. $\frac{1}{4} \times 4\frac{3}{4} = ?$ **6.** $\frac{1}{2} \times 7\frac{1}{4} = ?$ **9.** $\frac{3}{8} \times 2\frac{1}{2} = ?$

Check your answers on page 186.

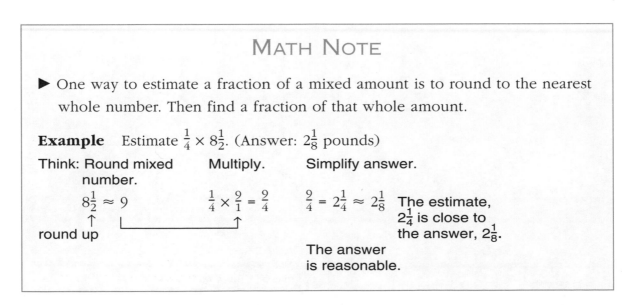

MATH NOTE

▶ One way to estimate a fraction of a mixed amount is to round to the nearest whole number. Then find a fraction of that whole amount.

Example   Estimate $\frac{1}{4} \times 8\frac{1}{2}$. (Answer: $2\frac{1}{8}$ pounds)

Think: Round mixed   Multiply.   Simplify answer.
   number.

$8\frac{1}{2} \approx 9$ $\frac{1}{4} \times \frac{9}{1} = \frac{9}{4}$ $\frac{9}{4} = 2\frac{1}{4} \approx 2\frac{1}{8}$ The estimate,
  ↑ $2\frac{1}{4}$ is close to
round up   the answer, $2\frac{1}{8}$.

           The answer
           is reasonable.

Dividing by a Common Fraction

The solution to a fraction division problem uses multiplication. The first step is to find the **reciprocal** of the fraction divisor. (A reciprocal is a fraction that when multiplied with another fraction equals 1.) The reciprocal is written in place of the divisor and multiplied by the other amount in the problem.

Example $2\frac{3}{4} \div \frac{1}{4} \rightarrow 2\frac{3}{4} \times \frac{4}{1}$ The reciprocal of $\frac{1}{4}$ is $\frac{4}{1}$. $\frac{1}{4} \times \frac{4}{1} = 1$

Exercise 13

Find the reciprocal of each fraction. (What fraction times the fraction equals 1?)

1. $\frac{1}{2}$ _____

2. $\frac{3}{4}$ _____

3. $\frac{5}{8}$ _____

4. $\frac{3}{10}$ _____

5. $\frac{7}{16}$ _____

Check your answers on page 186.

The Multiplication Problem

When you change the division problem to multiplication, keep the numbers in the same order. Write a **x** sign in place of the ÷ sign. Then follow the steps you learned to multiply fractions.

Example $2\frac{3}{4} \div \frac{1}{4} = ?$

Change the ÷ sign and write the reciprocal.

$2\frac{3}{4} \times \frac{4}{1}$

Rename and multiply.

$\frac{11}{4} \times \frac{4}{1} = \frac{44}{4}$ Rename $2\frac{3}{4}$ as an improper fraction.

Simplify answer.

$\frac{44}{4} = 11$

Exercise 14

Solve the equations. Simplify all answers. Show your work.

1. $9 \div \frac{1}{3} = ?$

2. $4\frac{2}{3} \div \frac{1}{2} = ?$

3. $8\frac{1}{2} \div \frac{1}{2} = ?$

4. $6\frac{3}{4} \div \frac{3}{4} = ?$

5. $6\frac{3}{4} \div \frac{2}{3} = ?$

6. $10 \div \frac{1}{2} = ?$

7. $7\frac{1}{3} \div \frac{1}{2} = ?$

8. $7 \div \frac{3}{4} = ?$

Check your answers on page 186.

Dividing by Whole or Mixed Numbers

When the divisor is a whole or a mixed number, first rename the number as an improper fraction. Then find the reciprocal of the improper fraction.

Example   Find the reciprocal of 4 and $3\frac{2}{3}$.

$4 = \frac{4}{1}$ $\frac{1}{4}$ is the reciprocal. $3\frac{2}{3} = \frac{11}{3}$ $\frac{3}{11}$ is the reciprocal.

Exercise 15

Rename the whole number or mixed number as an improper fraction. Then find its reciprocal. The first one is done for you.

1. $3\frac{1}{2} = \underline{\quad\frac{7}{2}\quad}$

reciprocal: $\underline{\quad\frac{2}{7}\quad}$

2. $4\frac{2}{3} = \underline{\qquad}$

reciprocal: $\underline{\qquad}$

3. $1\frac{5}{8} = \underline{\qquad}$

reciprocal: $\underline{\qquad}$

4. $8 = \underline{\qquad}$

reciprocal: $\underline{\qquad}$

5. $6\frac{1}{4} = \underline{\qquad}$

reciprocal: $\underline{\qquad}$

6. $15 = \underline{\qquad}$

reciprocal: $\underline{\qquad}$

Check your answers on page 186.

Solving the Problem

Remember: when changing a division problem to multiplication, keep the numbers in the same order. The reciprocal of the divisor should come after the × sign.

Example $20\frac{1}{2} \div 4$

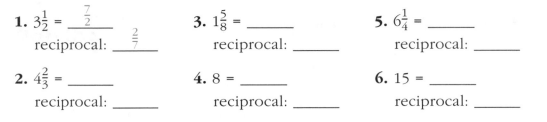

Change the ÷ sign and write the reciprocal.

$20\frac{1}{2} \times \frac{1}{4} = ?$

Rename $20\frac{1}{2}$ as an improper fraction and multiply.

$\frac{41}{2} \times \frac{1}{4} = \frac{41}{8}$

Simplify answer.

$\frac{41}{8} = 5\frac{1}{8}$

Exercise 16

Solve the equations. Simplify all answers. Show your work.

1. $2\frac{1}{2} \div 2 = ?$

2. $1\frac{3}{4} \div 4 = ?$

3. $4\frac{2}{3} \div 1\frac{1}{3} = ?$

4. $6\frac{1}{2} \div 1\frac{1}{2} = ?$

5. $\frac{3}{4} \div 2 = ?$

6. $\frac{1}{2} \div 2 = ?$

7. $8\frac{3}{4} \div 1\frac{1}{4} = ?$

8. $\frac{2}{3} \div 1\frac{1}{3} = ?$

Check your answers on page 186.

Fraction Checkup

How well did you understand Chapter 14?

1. The questions are based on the trail map. Read each question, then state which operation you would use to find the answer. Write *addition, subtraction, multiplication,* or *division.* **Do not solve.**

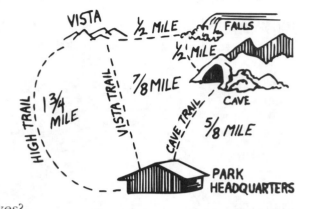

_____ **a.** From Headquarters, how far is it to the Falls by way of the Caves?

_____ **b.** What's the difference in mileage between the High Trail and the Vista Trail from the Vista to Park Headquarters?

_____ **c.** From Park Headquarters, what is the total mileage to walk up the Cave Trail to the Vista and return on the High Trail?

2. Solve the equations. Simplify all answers. Show your work.

a. $\frac{1}{8} + \frac{5}{8} = ?$ **d.** $1 - \frac{1}{4} = ?$ **g.** $\frac{1}{2} \times \frac{3}{4} = ?$ **j.** $\frac{3}{4} \div \frac{1}{2} = ?$

b. $\frac{2}{3} + \frac{3}{5} = ?$ **e.** $\frac{5}{8} - \frac{1}{2} = ?$ **h.** $\frac{1}{2} \times 2\frac{1}{2} = ?$ **k.** $5 \div \frac{1}{2} = ?$

c. $1\frac{1}{2} + 2\frac{3}{4} = ?$ **f.** $6\frac{1}{3} - 2\frac{2}{3} = ?$ **i.** $\frac{1}{2} \times 8 = ?$ **l.** $3\frac{1}{4} \div 2 = ?$

Check your answers on page 186.

POINTS TO REMEMBER

▶ To add and subtract fractions, the fractions must have the same denominator. Change unlike denominators to a common denominator.

▶ To multiply fractions, multiply numerators and denominators.

▶ To divide fractions, rename the division problem as a multiplication problem. The divisor is changed to its reciprocal.

UNIT 5
RATIOS AND PERCENTS

IN THIS UNIT, YOU WILL LEARN BASIC CONCEPTS AND PROBLEM SOLVING WITH RATIOS AND PERCENTS. YOU'LL LEARN HOW TO

▶ compare amounts using ratios and percents

▶ describe what ratios and proportions are

▶ state what amount a percent is equivalent to

▶ rename percents as common fractions and decimal fractions

▶ use a proportion to solve for an unknown amount

▶ solve for a percent of an amount

CHAPTER 15 | RATIO POWER

You can use two math operations to make comparisons of two amounts: subtraction or division.

For example, suppose a pair of gloves is on sale for five dollars. Its regular price is ten dollars. With subtraction, you can compare the difference in prices. How much do you save? Write the difference below.

The difference between $10 (regular price) and $5 (sale price)

$$\$10 - \$5 = \underline{\quad \$5 \quad}$$

regular price ↗   ↖ difference

Ratios use *division* to compare how many times larger or smaller one amount is than the other. For example, to show how many times smaller the sale price is than the regular price, use a ratio: $5 to $10 or 5:10.

Ratios can be expressed as fractions: $\frac{5}{10} = \frac{1}{2}$. The sale price is $\frac{1}{2}$ the original price.

Ratios can also be used to express a **rate** — the comparison of an amount to one unit of something. Examples:

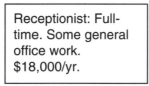

Receptionist: Full-time. Some general office work. $18,000/yr.

The ratio of an annual salary: $18,000 to 1 year

SPEED LIMIT 55

The ratio of a speed limit: 55 miles to 1 hour

In this chapter, you'll learn how to express a ratio and equivalent ratios as common fractions. And you'll learn how to use equivalent ratios to solve for an unknown amount.

TALK MATH

Do these activities with a partner or group.

1. Describe some situations in which you compared amounts by using division.

2. Take turns naming a fraction and discussing what is being compared. Example: You might say, "$\frac{3}{4}$." Your partner should say, "Three equal parts out of four equal parts."

Rates

Only one comparison is being made when ratios express a rate. When the second number isn't shown in a rate, that means the number is 1.

Example   Many businesses like entry-level clerks to type **45 words per minute**. The ratio of words per minute is **45** to **1**.

Remember: *per* means one.

Exercise 1

Write the ratio for each rate.

1. The ratio for 15 miles per hour:

_____ to _____

2. The ratio for 20 persons per table:

_____ to _____

3. The ratio of 8 servings per pizza:

_____ to _____

4. The ratio of 37 miles per gallon:

_____ to _____

Check your answers on page 187.

Fractions

Fractions — decimal fractions and common fractions — are ratios. Why is that?
Fractions make a comparison of two amounts: a part out of a total amount.

Example   A store advertised $\frac{1}{2}$ off (or .5 off) on all red-tag items. The ratio of $\frac{1}{2}$ is 1 to 2 halves. The ratio of .5 is 5 to 10 tenths.

Exercise 2

Write the ratio for each decimal or common fraction.

1. The ratio for .25:

_____ to _____

2. The ratio for .75:

_____ to _____

3. The ratio for $\frac{1}{4}$:

_____ to _____

4. The ratio for $\frac{3}{4}$:

_____ to _____

5. The ratio for $\frac{4}{5}$:

_____ to _____

6. The ratio for .8:

_____ to _____

Check your answers on page 187.

Expressing a Ratio

Remember, a ratio is the comparison of two amounts by division. You've learned to write a division expression using either a ÷ or $\overline{)\quad}$ sign. You can also write a division expression in the form of a common fraction. In a common fraction, the numerator (**Think:** dividend) is divided by the denominator (**Think:** divisor).

Example Harvey gets paid **$2** for every **$3** Jean earns.

$$3\overline{)2} \qquad 2 \div 3 \qquad \frac{2}{3} \begin{array}{l} \leftarrow \text{ the dividend} \\ \leftarrow \text{ the divisor} \end{array}$$

In the Form of a Common Fraction

When you express ratios as common fractions, the term being compared is written in the numerator's place. The term to which the first number is compared is written in the denominator's place.

Example Twenty-four persons were asked: "Do you favor unpaid leave up to six weeks for family emergencies?" **16** said yes; **8** said no. What is the ratio of yes to no answers?

$$\frac{16}{8} \begin{array}{l} \leftarrow \text{ yes answers} \\ \leftarrow \text{ no answers} \end{array}$$

To read a ratio, first say the amount being compared. Next say the word *to*. Then say the amount to which the first amount is compared: yes to no — 16 to 8.

Note: *never* change a ratio to a mixed number.

Exercise 3

Express each ratio in the form of a common fraction. The first one is done for you.

1. 2 to 1 $\frac{2}{1}$

2. 3 to 4 _____

3. 5 to 10 _____

4. 1 to 2 _____

5. 3 to 6 _____

6. 6 to 3 _____

7. 8 to 5 _____

8. 1 to 18 _____

9. 4 to 3 _____

10. 5 to 8 _____

11. 18 to 1 _____

12. 10 to 5 _____

Check your answers on page 187.

Reducing Ratios

Ratios usually are reduced to their lowest terms. You reduce the terms the same way you reduce the terms of a common fraction (as shown on page 132).

Example   By the end of the baseball season, a player got **96 hits** out of **144 times at bat**. What was his ratio of hits to times at bat?

$$\frac{96}{144} \quad \frac{96 \div 2 = 48 \div 6 = 8 \div 4 = 2}{144 \div 2 = 72 \div 6 = 12 \div 4 = 3}$$ The player averaged 2 hits for every 3 times at bat.

Notice how in the example the terms are divided until they can no longer be divided evenly by the same number. The ratio $\frac{96}{144}$ reduced to its lowest terms is $\frac{2}{3}$.

MATH NOTE

► Reduce ratios that are improper fractions to their lowest terms. Do not change ratios into mixed numbers. $(\frac{10}{5} = \frac{2}{1}$ or 10:5 or 2:1)

Exercise 4

Reduce the ratios to their lowest terms. Show all your work.

1. $\frac{4}{8}$ =   **3.** $\frac{12}{4}$ =   **5.** $\frac{20}{40}$ =   **7.** $\frac{45}{60}$ =   **9.** $\frac{36}{12}$ =

2. $\frac{6}{9}$ =   **4.** $\frac{20}{10}$ =   **6.** $\frac{24}{48}$ =   **8.** $\frac{56}{48}$ =   **10.** $\frac{120}{100}$ =

Exercise 5

Write each ratio in fraction form. Reduce ratios to lowest terms.

_____ **1.** 4 bottles of spaghetti sauce sell for $2. What is the ratio of *items* to *cost?*

_____ **2.** 15 out of 25 board members voted no on an important issue. What is the ratio of *members who voted no* to *total* members?

_____ **3.** A car travels 180 miles on 12 gallons. What is the ratio of *miles* to *gallons?*

_____ **4.** 200 invitations were sent out for a fund-raiser. 160 replies were received. What is the ratio of *replies* to total *invitations?*

Check your answers on page 187.

Equivalent Ratios

Two ratios are equivalent if they both compare the same amount. In the example, both rectangles equal 1 whole. The ratio of shaded parts in both rectangles to the whole is the same comparison.

Example   The ratio 6 to 10 is the same as 3 to 5 OR $\frac{6}{10} = \frac{3}{5}$.

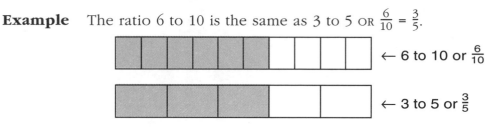

← 6 to 10 or $\frac{6}{10}$

← 3 to 5 or $\frac{3}{5}$

Proportions

Two equivalent ratios are called a **proportion**. Express a proportion by writing an equals sign (=) between the equivalent ratios. You read a proportion from left to right.

Example $\frac{3}{5} = \frac{6}{10}$ Think: 3 to 5 *is the same as* 6 to 10.

Exercise 6

Read each statement. Then write the proportion.

1. 2 to 4 is the same as 1 to 2.

_____

2. 6 to 2 is the same as 3 to 1.

_____

3. 10 to 20 is the same as 1 to 2.

_____

4. 14 to 10 is the same as 7 to 5.

_____

5. 25 to 15 is the same as 5 to 3.

_____

6. 32 to 24 is the same as 4 to 3.

_____

Check your answers on page 187.

A True Proportion

A proportion has two **cross products**. A cross product is the product of the numerator of one ratio times the denominator of the other ratio.

Example $\dfrac{2}{5} = \dfrac{4}{10}$ ← cross product: 4 x 5 = 20

cross product: 2 x 10 = 20

Exercise 7

Set up and find the cross products for each proportion.

1. $\dfrac{2}{3} = \dfrac{8}{12}$

   a. _____

   b. _____

2. $\dfrac{1}{4} = \dfrac{6}{24}$

   a. _____

   b. _____

3. $\dfrac{1}{6} = \dfrac{20}{120}$

   a. _____

   b. _____

4. $\dfrac{7}{10} = \dfrac{70}{100}$

   a. _____

   b. _____

Check your answers on page 187.

Cross Products Should Be Equal

In a true proportion, the cross products should be equal. If they aren't equal, the ratios are not equivalent.

Example $\dfrac{2}{5} \overset{?}{=} \dfrac{4}{10}$ $2 \times 10 \overset{?}{=} 4 \times 5$ $\dfrac{2}{5} \overset{?}{=} \dfrac{6}{9}$ $2 \times 9 \overset{?}{=} 6 \times 5$

$20 = 20$ $18 \neq 30$

true proportion   not a true proportion

Exercise 8

Prove that each expression is a true proportion. Write _true_ if it is a proportion. Write _false_ if it is not. The first one is done for you.

1. $\dfrac{1}{2} = \dfrac{12}{24}$

   $1 \times 24 = 24$
   $2 \times 12 = 24$
   true

2. $\dfrac{1}{4} = \dfrac{2}{8}$

3. $\dfrac{1}{4} = \dfrac{5}{12}$

4. $\dfrac{2}{3} = \dfrac{8}{12}$

5. $\dfrac{3}{5} = \dfrac{9}{15}$

6. $\dfrac{3}{4} = \dfrac{12}{20}$

7. $\dfrac{1}{3} = \dfrac{10}{30}$

8. $\dfrac{1}{2} = \dfrac{40}{120}$

9. $\dfrac{3}{8} = \dfrac{7}{16}$

10. $\dfrac{1}{2} = \dfrac{36}{72}$

Check your answers on page 187.

Solving for an Unknown

You can use proportions to help you solve for an unknown amount. The proportion will have three known amounts and one unknown amount. Write a variable for the unknown amount. Keep the terms being compared in the same order.

Example In a cookie recipe, the ratio of sugar to flour is **1 cup** to **2 cups**. If you use **5 cups** of flour, how much sugar should you use?

$$\frac{sugar}{flour} = \frac{sugar}{flour} \qquad \frac{1}{2} = \frac{s}{5}$$

Step 1. Find the cross product of the two known terms.

$$1 \times 5 = s \times 2$$
$$5 = s \times 2$$

Step 2. Divide the cross product by the third known term.

$$\frac{5}{2} = s$$
$$2\frac{1}{2} = s$$

The unknown amount is $2\frac{1}{2}$. For 5 cups of flour, you would use $2\frac{1}{2}$ **cups** of sugar.

To check your answer, find the cross product. It should equal the other cross product. Does 1×5 equal $2 \times 2\frac{1}{2}$? If so, it's a true proportion.

Exercise 9

Solve for the unknown amount. You can use your calculator to do the math.

1. $\frac{3}{5} = \frac{u}{15}$ u: _____

2. $\frac{1}{2} = \frac{4}{s}$ s: _____

3. $\frac{x}{3} = \frac{9}{27}$ x: _____

4. $\frac{g}{5} = \frac{6}{15}$ g: _____

5. $\frac{4}{a} = \frac{16}{20}$ a: _____

6. $\frac{1.5}{3} = \frac{r}{6}$ r: _____

Exercise 10

Set up a proportion to solve for the unknown amount. Let the letter *x* stand for the variable. Solve for *x*.

1. Suppose you earn $15 for each hour of overtime. How much do you make if you work 4.5 hours?

   a. proportion: _____

   b. x: $ _____

2. Suppose a car can travel 225 miles on 10 gallons of gasoline. How many miles does the car get on 1 gallon?

   a. proportion: _____

   b. x: _____ miles per gallon

Check your answers on page 187.

Ratio Checkup

How well did you understand Chapter 15?

1. Circle the comparison that is *not* a ratio. Explain why it is not a ratio.

    a. 6 out of 15 or $\frac{6}{15}$ **b.** 15 to 6 or $\frac{15}{6}$ **c.** $15 - 6 = 9$

_____

2. Write a ratio to express the amounts being compared. Use the common fraction form. Reduce ratios to their lowest terms.

    a. Ratio: _____   **b.** Ratio: _____   **c.** Ratio: _____

3. Prove that the proportions are true. Set up and find the cross products. If the ratios are equivalent, write *yes*. If they aren't, write *no*.

    a. $\frac{2}{3} = \frac{6}{9}$ **b.** $\frac{4}{5} = \frac{20}{25}$ **c.** $\frac{7}{8} = \frac{49}{56}$

    _____   _____   _____

4. Read each situation. Does the proportion show the situation? If not, write the correct proportion. Solve for the unknown amount.

    a. Suppose you commute to work. It's a round-trip of 90 miles per day. How many miles do you travel in a 6-day workweek? $x =$ _____

        $\frac{90}{1} = \frac{x}{6}$

    b. An optical shop is having a sale on eyeglasses: 2 pairs of eyeglasses for $89. How much is 1 pair of eyeglasses at that price? $x =$ _____

        $\frac{2}{89} = \frac{x}{1}$

Check your answers on page 187.

POINTS TO REMEMBER

▶ When you compare two amounts by division, you get a ratio.

▶ Fractions — decimal fractions and common fractions — are ratios. A ratio usually is expressed in its lowest terms.

▶ Two ratios that compare the same amount are equivalent. Equivalent ratios are called a *proportion*. The cross products should be equal.

CHAPTER 16 | PERCENT POWER

Percents. We read and hear about them everywhere. They're used to make comparisons of a part out of a whole. Here is just one example.

Sale advertisements for stores:

Percents are another form of fractions. Percents differ from decimal fractions and common fractions in this way: *percents represent actual amounts.*

The example below shows how a total amount and its parts can be expressed as percents.

Example A city council has 10 members.

100% of the city council represents ___10___ members.

60% of the city council represents ___6___ members.

50% of the city council represents ___5___ members.

10% of the city council represents ___1___ member.

In this chapter, you'll learn how to express percents. You'll also learn several ways to find the amount that a percent represents.

TALK MATH

Do these activities with a partner or group.

1. Describe some situations in which you've read or heard percents being used.

2. Take turns giving a decimal or common fraction and saying its ratio. For example, you might say, ".50." Your partner should say, "50 to 100."

Hundredths

Unlike common fractions and decimal fractions, percents have only one denominator. It is *hundredths*, or 1 whole divided into 100 equal parts. (The word *percent* means "for every 100.") 100 percent always represents 1 whole.

Example   40 percent of each shape is shaded. 60 percent is unshaded.

$$\frac{4}{10} = \frac{40}{100} = 40 \text{ percent}$$
$$\frac{6}{10} = \frac{60}{100} = 60 \text{ percent}$$

Exercise 1

The shapes below each represent 1 whole or 100 percent. What percent is shaded? What percent is unshaded?

1. a. _____ percent is shaded.

 b. _____ percent is unshaded.

2. a. _____ percent is shaded.

 b. _____ percent is unshaded.

Check your answers on page 188.

Expressing a Percent

As with decimal fractions, only the numerator is shown in a percent. The denominator is represented by the **percent sign**, %.

Example   40%   Read as "forty percent."

Exercise 2

Circle the percent in each sentence. Then write an expression for it.

_____ **1.** A computer store is having a 25 percent sale on all printers.

_____ **2.** Waiters hope that diners will leave at least a 15 percent tip.

_____ **3.** Some interest rates on mortgages have dropped to about 7 percent.

_____ **4.** Only 18 percent of all 18-year-olds at a city college are registered voters.

_____ **5.** Forty-two percent of a candy bar's 225 calories is fat.

Check your answers on page 188.

Percents Are Parts of a Whole

Percents are often used to show breakdowns of a *total*, or whole, amount. Often, breakdowns are shown on circle graphs such as the one in the example. The circle itself represents 100% or the total amount. The segments of the circle represent the percents, or parts, of the total amount. All the percents should add up to 100%.

Read the example. What amount is 100%? What amounts are 40% and 60%?

Example   In a survey of 20 people, 8 persons or 40% were Democrats, 12 or 60% were Republicans.

100% of the survey is 20 people. 8 people are 40%, and 12 people are 60%. Notice that the two percents, 40% and 60%, add up to 100%.

Exercise 3

Study each circle graph. Then answer the questions about it.

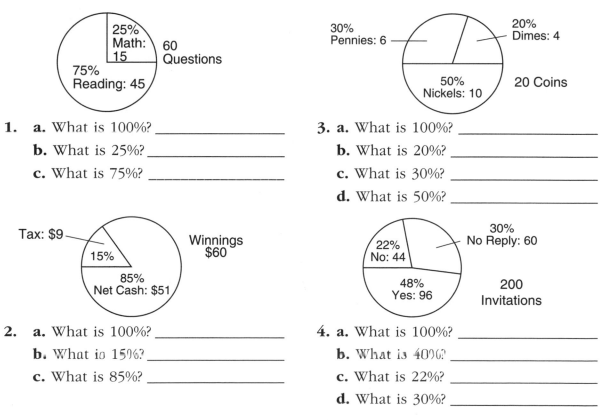

1. **a.** What is 100%? _____

    b. What is 25%? _____

    c. What is 75%? _____

3. **a.** What is 100%? _____

    b. What is 20%? _____

    c. What is 30%? _____

    d. What is 50%? _____

2. **a.** What is 100%? _____

    b. What is 15%? _____

    c. What is 85%? _____

4. **a.** What is 100%? _____

    b. What is 40%? _____

    c. What is 22%? _____

    d. What is 30%? _____

Check your answers on page 188.

Percents as Ratios

Percents can be expressed as ratios. Why do you think that is? Remember: fractions are ratios. How would you express a percent as a ratio?

Example   Express 50% as a ratio. 50% is 50-*hundredths*. The ratio then is 50 to 100 or $\frac{50}{100}$. The ratio can then be reduced to lowest terms.

Exercise 4

Express the percents in the ratio form of a common fraction. Reduce to lowest terms.

1. 10% = _____   **3.** 42% = _____   **5.** 75% = _____   **7.** 60% = _____

2. 5% = _____   **4.** 25% = _____   **6.** 80% = _____   **8.** 93% = _____

Check your answers on page 188.

Proportions

A proportion can show what amount out of the total a percent represents. One ratio in the proportion is the percent in fraction form. The other ratio is the actual part out of the total. Cross multiply to check for a true proportion.

Example   The regular price for a pillow is $15. The store takes off 25% or $3.75.

$$\text{percent} \rightarrow \frac{25}{100} = \frac{3.75}{15} \begin{array}{l} \leftarrow \text{part of the total amount} \\ \leftarrow \text{total amount} \end{array}$$

always 100 percent →

Think: 25% or 25 to 100 is the same as 3.75 to 15.

Exercise 5

Write a proportion comparing the whole percent to its part.

1. A stereo system costs $349. The store takes 20% off the price, or $69.80.

2. Sales tax is 5%. A purchase totals $25.00, of which $1.25 was tax.

3. Out of 26 games, the Padres won 13 games. The percentage of wins is 50%.

4. 72 out of 240 employees are new hires. They make up 30% of the company.

Check your answers on page 188.

Solving for a Percent

One of the more common problems we solve with percents is this: finding the actual amount that a percent represents. One solution is to set up a proportion.

The proportion would have a ratio for the percent and a ratio for the part out of the total. You would write a variable in place of the unknown part.

Example   A radio costs **$96**. The store discounts it by **25%**. What is 25% of $96?

percent → $\dfrac{25}{100}$ = $\dfrac{\$}{96}$ ← unknown part of the total
always 100 percent →   ← total

Once you have the proportion, how do you solve for the unknown part?

Example $\dfrac{25}{100}$ = $\dfrac{\$}{96}$ Think: Find the cross product of the two known terms.
$25 \times 96 = 2{,}400$

Divide the product by the third known term.

$2{,}400 \div 100 = 24$

The answer: 25% of $96 is **$24**.

Exercise 6

The problems are based on the circle graph. Set up a proportion, using a question mark for the variable. Solve for the unknown part. You can use a calculator to do the math.

1. How many students are Latino immigrants?

2. How many students are Asian immigrants?

3. How many students are European immigrants?

4. How many students are Southeast Asian immigrants?

5. How many students are of other ethnic backgrounds?

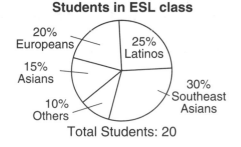

Ethnic Breakdown of Immigrant Students in ESL class

20% Europeans
25% Latinos
15% Asians
30% Southeast Asians
10% Others
Total Students: 20

Check your answers on page 188.

Percent = Decimal = Common Fraction
A percent, decimal and common fraction can represent the same amount.

Renaming Percents as Decimals

Percents can be renamed as decimal fractions. Most can be renamed without doing any math operation. Just remember this: percents have a denominator of *hundredths*. *Hundredths* in decimal fractions have two decimal places.

If the *hundredths* place in a decimal is filled with a 0 (zero), you can drop it. (For a review about dropping end zeros in decimal fractions, reread page 106.)

Example   Rename 50% as a decimal fraction.

drop end zero   50% = .50 = .5   Think: 50% is 50-*hundredths*; .50 = .5

Exercise 7
. . . .

Rename the percents as decimal fractions. Simplify the decimals.

1. 10% = _____   **3.** 42% = _____   **5.** 75% = _____

2. 5% = _____   **4.** 25% = _____   **6.** 80% = _____

Check your answers on page 188.

Renaming Percents as Common Fractions

To rename percents as common fractions, show the numerator with the denominator. Then reduce to the lowest terms.

Example   Rename 50% as a common fraction.

$$50\% = \frac{50}{100} \quad \frac{50 \div 50}{100 \div 50} = \frac{1}{2}$$

Exercise 8
. . . .

Rename the percents as common fractions. Reduce all fractions to their lowest terms. Show all your work.

1. 10% = _____   **3.** 42% = _____   **5.** 75% = _____

2. 5% = _____   **4.** 25% = _____   **6.** 80% = _____

Check your answers on page 188.

Other Solutions

Here's another way to find the actual amount that a percent represents. First rename the percent as a decimal or common fraction, then set up a multiplication problem. (For a review of multiplying by a decimal fraction, reread page 111; and by a common fraction, reread pages 144–146.)

Example Find 15% of $45.

As a decimal fraction problem:

$$\begin{array}{r} 4\,5 \\ \times\ \ .1\,5 \\ \hline 6\,.\,7\,5 \end{array}$$

Note: 6.75 is equivalent to $6\frac{3}{4}$.

As a common fraction problem:

$$\frac{15}{100} \times \frac{45}{1} = \frac{675}{100} = 6\frac{75}{100} = 6\frac{3}{4}$$

Exercise 9

Read each situation. Set up both the decimal fraction and common fraction solutions. Do not solve the problems.

1. A job exam has 120 questions. To qualify, you need to score at least 70%. How many questions do you need to answer correctly?

   a. decimal solution: _____   **b.** fraction solution: _____

2. Suppose a restaurant tab is $28. How much would you leave for a 15% tip?

   a. decimal solution: _____   **b.** fraction solution: _____

3. A shoe store sells a pair of shoes for $84. A mail-order house sells the same pair of shoes at 20% of the store's price. How much do the shoes cost by mail order?

   a. decimal solution: _____   **b.** fraction solution: _____

4. Suppose you borrow $2,500 from a bank. It charges 10% interest. How much will you pay in interest?

   a. decimal solution: _____   **b.** fraction solution: _____

5. An airline's new fare rate is 60% of its old rate. The fare from San Francisco to Honolulu used to be $258. What is the new fare?

   a. decimal solution: _____   **b.** fraction solution: _____

6. A doctor tells her patient to reduce her calorie intake by 30%. If the patient usually eats 2,800 calories a day, how many calories should she cut out?

   a. decimal solution: _____   **b.** fraction solution: _____

Check your answers on page 188.

Percent Checkup

How well did you understand Chapter 16?

1. Circle the number that represents a percent: $\frac{1}{4}$ 25% .25

2. Circle the number(s) with a denominator of *hundredths*: $\frac{12}{100}$ 12% .12

3. Write 75% as a ratio. _____

4. Rename 60 percent as

   a. a decimal fraction: _____   **b.** a common fraction: _____

Questions 5–7 are based on the circle graph.

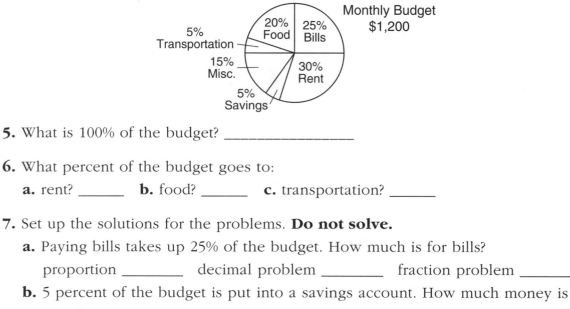

5. What is 100% of the budget? _____

6. What percent of the budget goes to:

   a. rent? _____   **b.** food? _____   **c.** transportation? _____

7. Set up the solutions for the problems. **Do not solve.**

   a. Paying bills takes up 25% of the budget. How much is for bills?

   proportion _____   decimal problem _____   fraction problem _____

   b. 5 percent of the budget is put into a savings account. How much money is saved?

   proportion _____   decimal problem _____   fraction problem _____

Check your answers on page 188.

POINTS TO REMEMBER

▶ Percents are fractions. They represent *hundredths* of 1 whole.

▶ 100% represents the total, or whole, amount.

▶ Percents are ratios. They compare an equal part out of a whole.

▶ You can use a proportion to solve for the unknown amount a percent represents.

▶ Percents can be renamed as decimal fractions and common fractions.

POST-TEST

Check your understanding of the problem-solving concepts and skills that you learned in this book.

Part A

Answer the questions.

1. Write the numbers that represent these amounts.
   _____ **a.** three *thousands* + five *hundreds* + zero *tens* + nine *ones*
   _____ **b.** six-*tenths* + eight-*hundredths* + three-*thousandths*
   _____ **c.** five and five-*tenths*
   _____ **d.** two-*thirds*
   _____ **e.** six and three-*fourths*

2. In the amount $42.16, on which side of the decimal point is:
   a. a whole number amount? _____   **b.** a fractional amount? _____

3. What is the place value of each digit in the number 873.406?
   a. 8: _____   **d.** 3: _____
   b. 6: _____   **e.** 7: _____
   c. 4: _____   **f.** 0: _____

4. Write the fractional amount 15-*hundredths* as a
   a. decimal fraction: _____ **b.** common fraction: _____ **c.** percent: ____

5. Simplify the following fractions:
   a. 2.50 = _____   **b.** .8000 = _____   **c.** $\frac{4}{8}$ = _____   **d.** $\frac{3}{2}$ = _____

6. Write the following comparisons as ratios in fraction form.
   a. a group of 15 women to 7 men: _____
   b. the cost of oat cereal: 2 boxes for $2.89: _____
   c. baseball player makes 7 hits out of 8 attempts: _____
   d. a store takes 25% off all its goods: _____

7. The following questions are based on the circle graph.
   a. What amount does 100% represent? _____ people
   b. What amount does 35% represent? _____ elderly
   c. Set up a proportion to show what amount 44% represents: _____

Los Altos: Population 200

21% Children: 42

44% Adults: 88

35% Elderly: 70

Part B

Unit 1: Whole Numbers

Solve the problems. Use a calculator or do the problem solving yourself.

1. $38 + 79 = $ _____   **3.** $2 \times 235 = $ _____   **5.** $378 \div 9 = $ _____

2. $72 - 15 = $ _____   **4.** $12 \times 44 = $ _____   **6.** $156 \div 12 = $ _____

Unit 2: Money

7. $\$156 + \$289 = $ _____   **9.** $\$5 - \$1.85 = $ _____   **11.** $\$21 \div 4 = $ _____

8. $\$13.14 - \$11.99 = $ _____   **10.** $12 \times \$4.25 = $ _____   **12.** $\$3.05 \div 5 = $ _____

Unit 3: Decimals

Simplify all decimal fractions.

13. $.8 + 2.45 + 1.076 = $ _____   **16.** $.5 \times 2.3 = $ _____   **19.** $.75 \div 5 = $ _____

14. $3 - 2.5 = $ _____   **17.** $1.2 \times 4.15 = $ _____   **20.** $4.6 \div .2 = $ _____

15. $4 \times 4.36 = $ _____   **18.** $1.3 \div 2 = $ _____   **21.** $24.6 \div 1.2 = $ _____

Unit 4: Fractions

Simplify improper fractions and reduce fractions to lowest terms.

22. $\frac{3}{8} + \frac{2}{8} = $ _____   **26.** $\frac{3}{4} - \frac{1}{2} = $ _____   **30.** $\frac{1}{2} \times 2\frac{1}{2} = $ _____

23. $\frac{1}{4} + \frac{2}{3} = $ _____   **27.** $3\frac{1}{3} - 1\frac{1}{2} = $ _____   **31.** $\frac{3}{4} \div \frac{1}{2} = $ _____

24. $2\frac{1}{2} + 1\frac{3}{4} = $ _____   **28.** $\frac{1}{2} \times \frac{3}{4} = $ _____   **32.** $3\frac{1}{2} \div 2 = $ _____

25. $\frac{2}{3} - \frac{1}{3} = $ _____   **29.** $\frac{1}{2} \times 4 = $ _____   **33.** $6\frac{1}{4} \div 2\frac{1}{2} = $ _____

Unit 5: Ratios and Percents

34. Find the unknown amount in the proportion: $\frac{2}{3} = \frac{x}{9}$

35. What is 10% of 12?

Part C

The following questions are based on the table. Show all your work.

1. How much less is Airline A's fare than Airline B?

2. How much is a round-trip ticket from Airline A?

One-way fares from San Francisco to Chicago
Airline A $284
Airline B $373

3. Airline A gives a 10% discount to senior citizens.
   a. How much is taken off the fare?

   b. How much is a one-way fare for a senior citizen?

Part D

The following questions are based on the grocery ad. Estimate the answers. Show all your work. You can use a calculator to get the exact answers.

Pine Market Holiday Sale

Young Turkey 78¢ pound

3 for $2 Vegetables

$1.29 Biscuits

$1.49 Potatoes, 15-pound bag

1. How much would a 10-pound turkey cost? _____

2. At the sale price, what is the price for 1 pound of potatoes? _____

3. How much would 1 package of vegetables cost? _____

4. If you use a 25¢-off coupon for the biscuits, how much will you pay? _____

5. Suppose you buy 1 bag of potatoes, 3 boxes of peas, 1 can of biscuits, and 1 turkey that costs $9.36.
   a. How much would you pay in all? _____

   b. If you pay with $20, how much change will you get? _____

Part E

Questions 1–5 are based on the recipe. Show all your work.

	Banana Bread	Yield: 2 loaves
Ingredients:		
$1\frac{3}{4}$ cups white flour	2 tsp baking soda	2 cups mashed bananas
$\frac{3}{4}$ cup whole wheat flour	1 cup liquid shortening	1 cup chopped nuts
$\frac{1}{4}$ cup oat flour	$1\frac{3}{4}$ cups sugar	$\frac{1}{2}$ cup raisins
$\frac{1}{2}$ tsp salt	3 eggs, slightly beaten	

1. How many *total* cups of flour are used for the recipe?

2. How much more white flour is used than oat flour?

3. If you double (2 times more) the recipe, how much oat flour will you need?

4. If you divided the recipe in half (by 2), how much sugar would you need?

5. The ratio of bananas per recipe is $\frac{2}{1}$. How many cups of mashed bananas would you need to make the recipe three times?

Part F

The following questions are about the sales ad. Show all your work. Round answers to the nearest penny.

1. What are the savings on the dishes?

Festive Floral
sale: $39.99
20 piece set
$80, regular price

2. What is the cost for 2 sets of dishes?

3. The set has 4 settings. What is the cost per setting?

4. Suppose you buy 1 set of Festive Floral on sale.
   a. If sales tax is 6%, how much will be added to the sale price?

   b. Including sales tax, what will be the total cost for 1 set?

Check your answers on page 172.

Post-Test Evaluation Chart

Use the answer key on page 172 to check your answers to the Post-Test. Then find the item number of each question you missed and circle it on the chart below. Next write the number of correct answers you had for each skill. If you need more practice in any skill, refer to the chapter that covers that skill area.

	Chapter	Skill	Item Numbers	Number Correct
Part A	1	Whole number place value	1	_____
	8	Place values in money	2	_____
	10	Understanding fractional amounts	3	_____
	13	Fractions	5	_____
	15	Ratios	6	_____
	16	Percents	4, 7	_____
Part B	3	Addition	1	_____
	4	Subtraction	2	_____
	6	Multiplication	3, 4	_____
	7	Division	5, 6	_____
	9	Money amounts	7, 8, 9, 10, 11, 12, 13, 14, 15	_____
	11	Decimals	16, 17, 18, 19, 20, 21, 22, 23, 24	_____
	14	Fractions/Mixed numbers	25, 26, 27, 28, 29, 30, 31, 32, 33, 34, 35, 36	_____
	15	Proportions	37	_____
	16	Percents	38	_____
Part C	9	Money amounts	1, 2, 3b	_____
	16	Percents	3a	_____
Part D	9	Money amounts	1, 2, 3, 4, 5a, 5b	_____
Part E	14	Fractions/Mixed numbers	1, 2, 3, 4	_____
	15	Ratios	5	_____
Part F	9	Money amounts	1, 2, 3, 4b	_____
	16	Percents	4a	_____

POST-TEST ANSWER KEY

Part A, page 167

1. **a.** 3,509
   b. .683
   c. 5.5
   d. $\frac{2}{3}$
   e. $6\frac{3}{4}$
2. **a.** left
   b. right
3. **a.** *hundreds*
   b. *thousandths*
   c. *tenths*
   d. *ones*
   e. *tens*
   f. *hundredths*
4. **a.** .15
   b. $\frac{15}{100}$
   c. 15%

5. **a.** 2.5
   b. .8
   c. $\frac{1}{2}$
   d. $1\frac{1}{2}$
6. **a.** $\frac{15}{7}$
   b. $\frac{2}{2.89}$
   c. $\frac{7}{8}$
   d. $\frac{25}{100}$
7. **a.** 200 people
   b. 70 elderly
   c. $\frac{44}{100} = \frac{88}{200}$

Part B, page 168

1. 117
2. 57
3. 470
4. 528
5. 42
6. 13
7. $445
8. $1.15
9. $3.15
10. $51.00
11. $5.25
12. $.61

13. 4.326
14. .5
15. 17.44
16. 1.15
17. 4.98
18. .65
19. .15
20. 23
21. 20.5
22. $\frac{5}{8}$
23. $\frac{11}{12}$
24. $4\frac{1}{4}$

25. $\frac{1}{3}$
26. $\frac{1}{4}$
27. $1\frac{5}{6}$
28. $\frac{3}{8}$
29. 2
30. $1\frac{1}{4}$
31. $1\frac{1}{2}$
32. $1\frac{3}{4}$
33. $2\frac{1}{2}$
34. $x = 6$
35. 1.20 or 1.2

Part C, page 169

1. $373 − $284 = $89
2. 2 × $284 = $568
3. **a.** solutions: .1 × 284 = _____ or $\frac{1}{10}$ × 284 = _____ or $\frac{10}{100} = \frac{x}{284}$; answer: $28.40
   b. $284 − $28.40 = $255.60

Part D, page 169

1. $.80 × 10 = $8; $7.80
2. $1.50 ÷ 15 = $.10; $.099 (rounded to nearest penny)
3. 2 ÷ 3 ≈ $.70; $.67 (rounded to nearest penny)
4. $1.30 − .25 = $1.05; $1.04
5. **a.** $1.50 + $2 + $1 + $9 = $13.50; $14.14
   b. $20 − 14 = $6; $5.86

Part E, page 170

1. $1\frac{3}{4} + \frac{3}{4} + \frac{1}{4} = 2\frac{3}{4}$ cups
2. $1\frac{3}{4} - \frac{1}{4} = 1\frac{2}{4} = 1\frac{1}{2}$ cups
3. $2 \times \frac{1}{4} = \frac{2}{4} = \frac{1}{2}$ cup
4. $1\frac{3}{4} \times \frac{1}{2} = \frac{7}{8}$ cup
5. $\frac{2}{1} = \frac{x}{3}$; $x = 6$ cups

Part F, page 170

1. $80 − $39.99 = $40.01
2. 2 × $39.99 = $79.98
3. $39.99 ÷ 4 = $10.00
4. **a.** .06 × 39.99 OR $\frac{3}{50}$ × 39.99 OR $\frac{6}{100} \times \frac{?}{39.99}$; answer: $2.40
   b. $39.99 + $2.40 = $42.39

ANSWER KEY

Unit 1: Whole Numbers

Chapter 1: Number Power

Exercise 1, page 3

1. a. 1 **b.** 6   **5. a.** 6 **b.** 3 **c.** 4
2. a. 8   **6. a.** 5 **b.** 9 **c.** 1
3. a. 2 **b.** 7   **7. a.** 9 **b.** 6 **c.** 3 **d.** 1
4. a. 0 **b.** 1 **c.** 4

Exercise 2, page 4

1. *ones*   **3.** *hundreds*
2. *tens*   **4.** *hundreds*

Exercise 3, page 5

1. *tens*   **6.** *tens*
2. *tens*   **7.** *ones*
3. *tens*   **8.** *thousands*
4. *hundreds*   **9.** *hundreds*
5. *ones*

Exercise 4, page 6

1. 30 *ones*   **5.** 40 *tens*
2. 50 *ones*   **6.** 70 *tens*
3. 80 *ones*   **7.** 90 *tens*
4. 120 *ones*   **8.** 150 *tens*

Exercise 5, page 6

1. 3 *hundreds*   **5.** 500 *ones*
2. 1 *hundred*   **6.** 800 *ones*
3. 4 *hundreds*   **7.** 300 *ones*
4. 6 *hundreds*   **8.** 900 *ones*

Exercise 6, page 7

1. 3 *tens* and 8 *ones*
2. 7 *ones*
3. 5 *tens* and 1 *one*
4. 1 *hundred* and 4 *tens* and 4 *ones*
5. 3 *hundreds* and 8 *tens* and 9 *ones*
6. 3 *thousands* and 5 *hundreds* and 2 *tens* and 5 *ones*

Exercise 7, page 8

1. 2 *tens* and 0 *ones*
2. 9 *tens* and 9 *ones*
3. 1 *ten* and 5 *ones*
4. 3 *tens* and 8 *ones*
5. 2 *hundreds* and 4 *tens*
6. 3 *hundreds* and 1 *ten*
7. 1 *hundred* and 9 *tens*

8. 1 *thousand* and 0 *hundreds*
9. 5 *thousands* and 6 *hundreds*
10. 7 *thousands* and 2 *hundreds*

Exercise 8, page 9

1. 20 *ones*   **6.** 90 *tens*
2. 70 *ones*   **7.** 10 *hundreds*
3. 40 *ones*   **8.** 40 *hundreds*
4. 30 *tens*   **9.** 150 *ones*
5. 50 *tens*   **10.** 120 *tens*

Exercise 9, page 10

1. 30; The *ones* digit is greater than 5.
2. 30; The *ones* digit is less than 5.
3. 70; The *ones* digit is 5.
4. 120; The *ones* digit is less than 5.
5. 160; The *ones* digit is greater than 5.

Exercise 10, page 10

1. 200; The *tens* digit is less than 5.
2. 200; The *tens* digit is greater than 5.
3. 500; The *tens* digit is 5.

Exercise 11, page 11

The answers will vary depending on the student's calculator. Possible answers are:
1. The display shows nothing or a zero or a decimal point.
2. 8 numbers; *ten millions*
3. The display shows nothing or a zero.

Exercise 12, page 12

The answers will appear on the calculator display.

Number Checkup, page 13

1. a. Zero helps a number hold its value by filling an empty number place.
   b. Because 10 in a number place equals 1 of the next higher place.
2. Three possible answers:
   250; 2 fills the *hundreds* place.
   520; 2 fills the *tens* place.
   502; 2 fills the *ones* place.
3. 2 *hundreds* 4 *ones* 3 *tens* 1 *thousands*
4. a. 60 **b.** 33 **c.** 109
5. a. 60 *ones* **b.** 4 *hundreds* and 5 *tens*
6. a. 10, 50, 390 **b.** 200, 400, 600
7. 3 5 8

Chapter 2: Addition and Subtraction Facts

Exercise 1, page 15

1. 1	6. 1
2. 0	7. 0
3. 2	8. 4
4. 0	9. 1
5. 3	10. 0

Exercise 2, page 15

1. 5	9. 7	17. 1
2. 2	10. 3	18. 0
3. 1	11. 2	19. 9
4. 0	12. 1	20. 4
5. 6	13. 0	21. 3
6. 2	14. 8	22. 2
7. 1	15. 3	23. 1
8. 0	16. 2	24. 0

Exercise 3, page 16

1. 4	9. 5
2. 3	10. 4
3. 2	11. 3
4. 1	12. 6
5. 5	13. 5
6. 4	14. 4
7. 3	15. 6
8. 2	16. 5

Exercise 4, page 16

1. 7	3. 7
2. 6	4. 8

Exercise 5, page 18

1. 9	15. 5	29. 1
2. 8	16. 4	30. 0
3. 7	17. 3	31. 9
4. 6	18. 2	32. 8
5. 5	19. 1	33. 7
6. 4	20. 0	34. 6
7. 3	21. 9	35. 5
8. 2	22. 8	36. 4
9. 1	23. 7	37. 3
10. 0	24. 6	38. 2
11. 9	25. 5	39. 1
12. 8	26. 4	40. 0
13. 7	27. 3	
14. 6	28. 2	

Exercise 6, page 18

1. 9	18. 2	35. 5
2. 8	19. 1	36. 4
3. 7	20. 0	37. 3
4. 6	21. 9	38. 2
5. 5	22. 8	39. 1
6. 4	23. 7	40. 0
7. 3	24. 6	41. 9
8. 2	25. 5	42. 8
9. 1	26. 4	43. 7
10. 0	27. 3	44. 6
11. 9	28. 2	45. 5
12. 8	29. 1	46. 4
13. 7	30. 0	47. 3
14. 6	31. 9	48. 2
15. 5	32. 8	49. 1
16. 4	33. 7	50. 0
17. 3	34. 6	

Facts Checkup, page 19

1. **a.** A number subtracted from the same number equals 0.
   b. Numbers can be added in any order, and the answer is the same.
2. **a.** 5 **b.** 7 **c.** 3 **d.** 8 **e.** 10 **f.** 13 **g.** 12 **h.** 11 **i.** 18 **j.** 15 **k.** 15 **l.** 14
3. **a.** 5 **b.** 3 **c.** 6 **d.** 1 **e.** 5 **f.** 4 **g.** 4 **h.** 9 **i.** 8 **j.** 9 **k.** 8 **l.** 7
4. **a.** 6; 6 − 1 = 5 and 6 − 5 = 1
   b. 7; 7 − 4 = 3 and 7 − 3 = 4 **c.** 13; 13 − 5 = 8 and 13 − 8 = 5 **d.** 11; 11 − 9 = 2 and 11 − 2 = 9
   e. 13; 13 − 6 = 7 and 13 − 7 = 6

Chapter 3: Addition Power

The numbers in Exercise 1 can be in either order.

Exercise 1, page 21

1.
```
   3 5
 + 1 8
```
2.
```
   4 4
 + 2 0
```
3.
```
  1 6 1
+ 4 6 5
```
4.
```
   2 3 0
 + 3 0 4
```
5.
```
   2 5 5
 +  9 9
```

Exercise 2, page 22

1. 29	5. 47
2. 96	6. 385
3. 30	7. 69
4. 467	8. 866

Exercise 3, page 22

1. 42	4. 396
2. 45	5. 893
3. 70	6. 899

Exercise 4, page 23

1. 33 + 42 = ?   **4.** 333 + 500 = ?
2. 51 + 18 = ?   **5.** 416 + 64 = ?
3. 125 + 125 = ?   **6.** 85 + 909 = ?

Exercise 5, page 24

1. 42   **6.** 221
2. 30   **7.** 496
3. 62   **8.** 440
4. 84   **9.** 647
5. 95

Exercise 6, page 25

1. 127   **6.** 140
2. 108   **7.** 107
3. 135   **8.** 189
4. 146   **9.** 127
5. 118   **10.** 139

Exercise 7, page 26

1. 39   **7.** 74
2. 49   **8.** 81
3. 51   **9.** 50
4. 70   **10.** 80
5. 72   **11.** 100
6. 74   **12.** 118

Exercise 8, page 27

1.
```
  10
  30
+ 10
  50
```
2.
```
  20
  20
+ 90
 130
```
3.
```
  90
 100
+100
 290
```
4.
```
  50
  10
+ 40
 100
```
5.
```
  50
  50
+400
 500
```
6.
```
 300
 300
+300
 900
```
7.
```
 200
 400
+200
 800
```
8.
```
  300
   80
+ 900
1,280
```
9.
```
   50
  600
+ 400
1,050
```
10.
```
  10
  40
  30
+ 20
 100
```
11.
```
  50
  20
  30
+ 40
 140
```

Exercise 9, page 28

1. a. estimate
2. a. estimate
3. b. exact
4. a. estimate
5. b. exact
6. a. estimate

Addition Checkup, page 29

1. a. You would add numbers to get a total amount.
    b. You could add 6 miles and 4 miles. The amounts have the same units of measurement.
2. First add the *ones* place: 9 + 4 = 13 *ones*. Regroup: 13 *ones* = 1 *ten* + 3 *ones*. Write 3 *ones*. Carry 1 *ten*. Add the *tens* place: 1 + 2 + 5 = 8 *tens*. Then add the *hundreds* place: 2 + 3 = 5 *hundreds*.

3. a.
```
  40
  10
+ 50
 100
```
b.
```
 100
  70
+200
 370
```
c.
```
  70
  30
+ 20
 120
```
d.
```
  50
  20
  20
+ 10
 100
```
4. a. estimate

Chapter 4: Subtraction Power

Exercise 1, page 31

1. 7   **5.** 224
2. 24   **6.** 100
3. 32   **7.** 24
4. 80   **8.** 321

Exercise 2, page 32

1. 39 − 28 = ?   **4.** 92 − 87 = ?
2. 49 − 42 = ?   **5.** 500 − 200 = ?
3. 56 − 16 = ?   **6.** 618 − 394 = ?

Exercise 3, page 33

1. 8   **7.** 64
2. 16   **8.** 47
3. 9   **9.** 19
4. 29   **10.** 29
5. 47   **11.** 45
6. 9   **12.** 78

12.
```
  100
   80
   20
+ 100
  300
```

Exercise 4, page 34

1. 58		**7.** 41
2. 63		**8.** 37
3. 149		**9.** 640
4. 173		**10.** 215
5. 183		**11.** 223
6. 216		**12.** 406

Exercise 5, page 35

1. 8	**6.** 17
2. 21	**7.** 65
3. 23	**8.** 55
4. 26	**9.** 154
5. 24	**10.** 301

Exercise 6, page 35

1.
$$80 - 20 = 60$$
Answer given is not reasonable; correct answer: 62.

2.
$$50 - 10 = 40$$
Answer given is not reasonable; correct answer: 45.

3.
$$60 - 50 = 10$$
Answer given is not reasonable; correct answer: 7.

4.
$$90 - 50 = 40$$
Answer given is reasonable.

5.
$$80 - 20 = 60$$
Answer given is reasonable.

6.
$$900 - 400 = 500$$
Answer given is reasonable.

7.
$$400 - 100 = 300$$
Answer given is reasonable.

8.
$$300 - 200 = 100$$
Answer given is not reasonable; correct answer: 108.

9.
$$800 - 200 = 600$$
Answer given is reasonable.

10.
$$500 - 200 = 300$$
Answer given is reasonable.

11.
$$800 - 300 = 500$$
Answer given is not reasonable; correct answer: 529.

12.
$$800 - 700 = 100$$
Answer given is not reasonable; correct answer: 129.

Exercise 7, page 37

1. solution: 55 − 30 = ____
estimate: 60 − 30 = 30
answer: 25 miles per hour

2. solution: 62 + 25 = ____
estimate: 60 + 30 = 90
answer: 87 miles

3. solution: 30 − 22 = ____
estimate: 30 − 20 = 10
answer: $8

4. solution: 28 − 19 = ____
estimate: 30 − 20 = 10
answer: 9 vacation days

Subtraction Checkup, page 38

1. a. To get the difference between two amounts
   b. To find out how much more is needed to reach a given amount
2. First subtract the *ones* place. 2 is smaller than 7, so need to regroup. Borrow 1 *ten*. That leaves 8 *tens*. Write 8 *tens*. Regroup 1 *ten* as 10 *ones* and add to *ones*: 10 + 2 = 12 *ones*. Now subtract *ones*: 12 − 7 = 5 *ones*. Subtract *tens*: 8 − 5 = 3 *tens*. Subtract *hundreds*: 6 − 3 = 3 *hundreds*.
3. a. 80 − 30 = 50 Answer given is not reasonable; correct answer: 45.
   b. 200 − 100 = 100 Answer given is reasonable.
   c. 300 − 50 = 250 Answer given is reasonable.
   d. 400 − 100 = 300 Answer given is not reasonable; correct answer: 322.
4. solution: 750 − 625 = ?; estimate: 800 − 600 = 200; answer: $125

Chapter 5: Multiplication and Division Facts

Exercise 1, page 40

1. 1		**13.** 8		**25.** 21
2. 2		**14.** 10		**26.** 24
3. 3		**15.** 12		**27.** 27
4. 4		**16.** 14		**28.** 4
5. 5		**17.** 16		**29.** 8
6. 6		**18.** 18		**30.** 12
7. 7		**19.** 3		**31.** 16
8. 8		**20.** 6		**32.** 20
9. 9		**21.** 9		**33.** 24
10. 2		**22.** 12		**34.** 28
11. 4		**23.** 15		**35.** 32
12. 6		**24.** 18		**36.** 36

Exercise 2, page 41

1. 5	11. 14	21. 48
2. 10	12. 21	22. 56
3. 15	13. 28	23. 9
4. 20	14. 35	24. 18
5. 6	15. 42	25. 27
6. 12	16. 8	26. 36
7. 18	17. 16	27. 45
8. 24	18. 24	28. 54
9. 30	19. 32	29. 63
10. 7	20. 40	30. 72

Exercise 3, page 43

1. 9	13. 6	25. 3
2. 8	14. 5	26. 2
3. 7	15. 4	27. 1
4. 6	16. 3	28. 9
5. 5	17. 2	29. 8
6. 4	18. 1	30. 7
7. 3	19. 9	31. 6
8. 2	20. 8	32. 5
9. 1	21. 7	33. 4
10. 9	22. 6	34. 3
11. 8	23. 5	35. 2
12. 7	24. 4	36. 1

Exercise 4, page 43

1. 9	16. 3	31. 6
2. 8	17. 2	32. 5
3. 7	18. 1	33. 4
4. 6	19. 9	34. 3
5. 5	20. 8	35. 2
6. 4	21. 7	36. 1
7. 3	22. 6	37. 9
8. 2	23. 5	38. 8
9. 1	24. 4	39. 7
10. 9	25. 3	40. 6
11. 8	26. 2	41. 5
12. 7	27. 1	42. 4
13. 6	28. 9	43. 3
14. 5	29. 8	44. 2
15. 4	30. 7	45. 1

Facts Checkup, page 44

1. **a.** A number times 0 is 0. **b.** 0 divided by a number is 0.
2. **a.** 9 **b.** 5 **c.** 6 **d.** 18 **e.** 9 **f.** 18 **g.** 7 **h.** 6 **i.** 27 **j.** 40 **k.** 63 **l.** 6 **m.** 81 **n.** 9 **o.** 56
3. **a.** 21; 21 ÷ 3 = 7 and 21 ÷ 7 = 3
   b. 10; 10 ÷ 5 = 2 and 10 ÷ 2 = 5
   c. 32; 32 ÷ 8 = 4 and 32 ÷ 4 = 8
   d. 27; 27 ÷ 3 = 9 and 27 ÷ 9 = 3
   e. 24; 24 ÷ 4 = 6 and 24 ÷ 6 = 4

Chapter 6: Multiplication Power

Exercise 1, page 46

1. 2 8 × 3	4. 5 3 0 × 7
2. 4 5 × 6	5. 2 5 6 × 1 0
3. 6 5 × 1 9	6. 7 0 9 × 1 2

Exercise 2, page 47

1. 39	4. 226
2. 84	5. 699
3. 88	6. 555

Exercise 3, page 47

1. 80	4. 603
2. 90	5. 800
3. 440	6. 840

Exercise 4, page 48

The multiplier can come before or after the multiplication sign.

1. 15 × 6 = ____	4. 56 × 10 = ____
2. 76 × 9 = ____	5. 5 × 249 = ____
3. 12 × 87 = ____	6. 854 × 24 = ____

Exercise 5, page 50

1. 2 × 2 = 4; 4 + 5 = 9	6. 2 × 2 = 4; 4 + 3 = 7
2. 3 × 1 = 3; 3 + 2 = 5	7. 4 × 2 = 8; 8 + 2 = 10
3. 2 × 4 = 8; 8 + 1 = 9	8. 2 × 3 = 6; 6 + 5 = 11
4. 3 × 2 = 6; 6 + 3 = 9	9. 5 × 0 = 0; 0 + 3 = 3
5. 1 × 6 = 6; 6 + 1 = 7	

Exercise 6, page 51

1. 38	7. 236
2. 70	8. 860
3. 81	9. 580
4. 92	10. 429
5. 96	11. 368
6. 96	12. 755

Exercise 7, page 52

1. 416	6. 1,525
2. 530	7. 4,381
3. 1,664	8. 1,227
4. 1,836	9. 4,848
5. 2,142	

Exercise 8, page 53

1. 150	**7.** 750
2. 280	**8.** 870
3. 320	**9.** 1,680
4. 280	**10.** 1,280
5. 880	**11.** 1,560
6. 930	**12.** 1,980

Exercise 9, page 54

1.
$$\begin{array}{r} 24 \\ \times\ \ 2 \\ \hline 48 \end{array} \qquad \begin{array}{r} 24 \\ \times\ 10 \\ \hline 240 \end{array} \qquad \begin{array}{r} 48 \\ +\ 240 \\ \hline 288 \end{array}$$

2.
$$\begin{array}{r} 32 \\ \times\ \ 3 \\ \hline 96 \end{array} \qquad \begin{array}{r} 32 \\ \times\ 10 \\ \hline 320 \end{array} \qquad \begin{array}{r} 1 \\ 96 \\ +\ 320 \\ \hline 416 \end{array}$$

3.
$$\begin{array}{r} 11 \\ \times\ \ 5 \\ \hline 55 \end{array} \qquad \begin{array}{r} 11 \\ \times\ 10 \\ \hline 110 \end{array} \qquad \begin{array}{r} 55 \\ +\ 110 \\ \hline 165 \end{array}$$

4.
$$\begin{array}{r} 21 \\ \times\ \ 4 \\ \hline 84 \end{array} \qquad \begin{array}{r} 21 \\ \times\ 20 \\ \hline 420 \end{array} \qquad \begin{array}{r} 1 \\ 84 \\ +\ 420 \\ \hline 504 \end{array}$$

5.
$$\begin{array}{r} 1 \\ 253 \\ \times\ \ \ 2 \\ \hline 506 \end{array} \qquad \begin{array}{r} 253 \\ \times\ \ \ 10 \\ \hline 2,530 \end{array} \qquad \begin{array}{r} 1 \\ 506 \\ +\ 2,530 \\ \hline 3,036 \end{array}$$

6.
$$\begin{array}{r} 48 \\ \times\ \ 1 \\ \hline 48 \end{array} \qquad \begin{array}{r} 48 \\ \times\ 20 \\ \hline 960 \end{array} \qquad \begin{array}{r} 1 \\ 48 \\ +\ \ 960 \\ \hline 1,008 \end{array}$$

Exercise 10, page 55

1.
$$\begin{array}{r} 40 \\ \times\ \ 4 \\ \hline 160 \end{array}$$
5.
$$\begin{array}{r} 100 \\ \times\ \ \ 3 \\ \hline 300 \end{array}$$

2.
$$\begin{array}{r} 40 \\ \times\ \ 6 \\ \hline 240 \end{array}$$
6.
$$\begin{array}{r} 200 \\ \times\ \ \ 2 \\ \hline 400 \end{array}$$

3.
$$\begin{array}{r} 90 \\ \times\ \ 9 \\ \hline 810 \end{array}$$
7.
$$\begin{array}{r} 200 \\ \times\ \ \ 4 \\ \hline 800 \end{array}$$

4.
$$\begin{array}{r} 60 \\ \times\ \ 7 \\ \hline 420 \end{array}$$
8.
$$\begin{array}{r} 300 \\ \times\ \ \ 5 \\ \hline 1,500 \end{array}$$

Exercise 11, page 55

1.
$$\begin{array}{r} 20 \\ \times\ \ 20 \\ \hline 400 \end{array}$$
exact: 270

5.
$$\begin{array}{r} 60 \\ \times\ \ 10 \\ \hline 600 \end{array}$$
exact: 672

2.
$$\begin{array}{r} 10 \\ \times\ \ 10 \\ \hline 100 \end{array}$$
exact: 168

6.
$$\begin{array}{r} 70 \\ \times\ \ 40 \\ \hline 2,800 \end{array}$$
exact: 2,592

3.
$$\begin{array}{r} 30 \\ \times\ \ 20 \\ \hline 600 \end{array}$$
exact: 450

7.
$$\begin{array}{r} 90 \\ \times\ \ 20 \\ \hline 1,800 \end{array}$$
exact: 1,602

4.
$$\begin{array}{r} 40 \\ \times\ \ 20 \\ \hline 800 \end{array}$$
exact: 864

8.
$$\begin{array}{r} 90 \\ \times\ \ 20 \\ \hline 1,800 \end{array}$$
exact: 1,840

Exercise 12, page 57

1. $3 \times 425 =$ ____; $3 \times 400 = 1,200$ OR $3 \times 430 = 1,290$; $1,275

2. $4 \times 28 =$ ____; $4 \times 30 = 120$; 112 calories

3. $101 + 275 =$ ____; $100 + 300 = 400$ OR $100 + 280 = 380$; $376

4. $5 \times 75 =$ ____; $5 \times 80 = 400$; 375 miles

Multiplication Checkup, page 58

1. a. Both operations are used to get a total amount.
b. Addition combines two or more amounts that are alike. Multiplication combines one amount a given number of times.

2. First, multiply 4×0 *ones* = 0. Write 0 *ones*. Then multiply 4×5 *tens* = 20 *tens*. Regroup 20 as 2 *hundreds* and 0 *tens*. Write 0 *tens* and carry 2 *hundreds*. 4×2 *hundreds* = 8 *hundreds*. Add 8 + 2 = 10 *hundreds*. Regroup 10 *hundreds* as 1 *thousand*. Write 0 *hundreds*. Carry 1 *thousand*. Write 1 *thousand*.

3. a. $2 \times 40 = 80$ **b.** $3 \times 500 = 1,500$ **c.** $3 \times 320 = 960$ or $3 \times 300 = 900$ **d.** $10 \times 30 = 300$

4. $3 \times 48 =$ ____; $3 \times 50 = 150$; $144

Chapter 7: Division Power

Exercise 1, page 60

1. $2\overline{)44}$ **4.** $2\overline{)248}$
2. $6\overline{)39}$ **5.** $10\overline{)380}$
3. $4\overline{)290}$ **6.** $32\overline{)96}$

Exercise 2, page 61

1.
```
    1 3
2 ) 2 6
  - 2 x
    0 6
    - 6
      0
```

2.
```
    2 1
2 ) 4 2
  - 4 x
    0 2
    - 2
      0
```

3.
```
    2 1
3 ) 6 3
  - 6 x
    0 3
    - 3
      0
```

4.
```
    1 1
6 ) 6 6
  - 6 x
    0 6
    - 6
      0
```

5.
```
    4 4
2 ) 8 8
  - 8 x
    0 8
    - 8
      0
```

6.
```
    4 2
2 ) 8 4
  - 8 x
    0 4
    - 4
      0
```

7.
```
    3 1
3 ) 9 3
  - 9 x
    0 3
    - 3
      0
```

8.
```
    2 1
4 ) 8 4
  - 8 x
    0 4
    - 4
      0
```

Exercise 3, page 62

1. $50 \div 5 = x$
2. $69 \div 3 = x$
3. $168 \div 8 = x$
4. $382 \div 2 = x$
5. $84 \div 12 = x$
6. $520 \div 14 = x$

Exercise 4, page 63

1. 121
2. 112
3. 114
4. 111
5. 314
6. 231
7. 441
8. 333

Exercise 5, page 65

1. 43
2. 42
3. 92
4. 50
5. 31
6. 72
7. 60
8. 61
9. 51
10. 81
11. 105
12. 91

Exercise 6, page 66

1. 109
2. 105
3. 106
4. 107
5. 207
6. 207
7. 108
8. 105
9. 408
10. 309
11. 109
12. 109

Exercise 7, page 67

1.
```
    1 5
2 ) 3 0
  - 2 x
    1 0
  - 1 0
      0
```

2.
```
    2 6
3 ) 7 8
  - 6 x
    1 8
  - 1 8
      0
```

3.
```
    1 8
3 ) 5 4
  - 3 x
    2 4
  - 2 4
      0
```

4.
```
    3 8
2 ) 7 6
  - 6 x
    1 6
  - 1 6
      0
```

5.
```
    1 2 7
2 ) 2 5 4
  - 2 x
    0 5
  - 4 x
    1 4
  - 1 4
      0
```

6.
```
    2 1 8
2 ) 4 3 6
  - 4 x
    0 3
  - 2 x
    1 6
  - 1 6
      0
```

7.
```
    1 5
6 ) 9 0
  - 6 x
    3 0
  - 3 0
      0
```

8.
```
    2 5 1
3 ) 7 5 3
  - 6 x
    1 5
  - 1 5
    0 3
  - 3
      0
```

Exercise 8, page 68

1. 4
2. 5
3. 3
4. 2
5. 4
6. 4
7. 13
8. 9
9. 3

Exercise 9, page 69

1. 18
2. 12
3. 20
4. 25
5. 32
6. 28
7. 24
8. 18
9. 30

Exercise 10, page 70

Estimates will vary, depending on the compatible numbers chosen.

1. $37 \approx 36; 36 \div 6 = 6$
2. $50 \approx 49; 49 \div 7 = 7$
3. $65 \approx 63; 63 \div 9 = 7$
4. $68 \approx 64; 64 \div 8 = 8$ OR
$68 \approx 72; 72 \div 8 = 9$
5. $28 \approx 27; 27 \div 3 = 9$
6. $46 \approx 45; 45 \div 5 = 9$
7. $71 \approx 72; 72 \div 9 = 8$
8. $31 \approx 32; 32 \div 4 = 8$
9. $52 \approx 48; 48 \div 8 = 6$ OR
$52 \approx 56; 56 \div 8 = 7$

Exercise 11, page 71

Estimates will vary, depending on the compatible numbers chosen.

1. $57 \approx 60; 60 \div 2 = 30$
2. $49 \approx 45; 45 \div 3 = 15$
3. $93 \approx 100; 100 \div 4 = 25$
4. $246 \approx 240; 240 \div 6 = 40$
5. $302 \approx 320; 320 \div 8 = 40$
6. $153 \approx 150; 150 \div 5 = 30$
7. $207 \approx 200; 200 \div 5 = 40$
8. $318 \approx 320; 320 \div 4 = 80$
9. $486 \approx 480; 480 \div 8 = 60$

Exercise 12, page 71

Estimates will vary, depending on the compatible numbers chosen.

1. $1,680 \approx 1,600$, and $39 \approx 40; 1,600 \div 40 = 40$
2. $1,808 \approx 1,800$, and $17 \approx 18; 1,800 \div 18 = 100$
3. $252 \approx 260$, and $14 \approx 13; 260 \div 13 = 20$
4. $11 \approx 10; 1,980 \div 10 = 198$
5. $6,212 \approx 6,000; 6,000 \div 12 = 500$
6. $16,589 \approx 16,000$, and $15 \approx 16; 16,000 \div 16 = 1,000$
7. $3,342 \approx 3,300$, and $12 \approx 11; 3,300 \div 11 = 300$
8. $6,280 \approx 6,300$, and $92 \approx 90; 6,300 \div 90 = 70$
9. $4,560 \approx 4,500$, and $16 \approx 15; 4,500 \div 15 = 300$

Exercise 13, page 72

1. $310 \div 5 = \underline{\quad}; 300 \div 5 = 60;$ 62 words per minute
2. $60 \div 12 = \underline{\quad}; 60 \div 10 = 6;$ 5 years
3. $156 - 45 = \underline{\quad}; 160 - 50 = 110;$ 111 pounds
4. $324 \div 4 = \underline{\quad}; 320 \div 4 = 80;$ 81 calories

Division Checkup; page 73

1. **a.** Division and subtraction are alike because they separate amounts.
   b. Subtraction separates two like amounts to get a remainder. Division separates an amount into equal groups and solves for the amount in each group.
2. 1 *hundred* $\div 2$ can't be done. Regroup 1 *hundred* as 10 *tens* and add to *tens* place: $10 + 4 = 14$ *tens*. Divide into *tens*: $14 \div 2 = 7$ *tens*; $2 \times 7 = 14; 14 - 14 = 0$. Divide into *ones*: $6 \div 2 = 3$ *ones*; $2 \times 3 = 6; 6 - 6 = 0$.
3. **a.** estimate: 100; answer: 102 **b.** estimate: 90; answer: 93 **c.** estimate: 70; answer: 76 **d.** estimate: 200; answer: 210
4. $588 \div 28 = \underline{\quad};$ estimate: 20; answer: 21 miles per gallon

Unit 2: Money

Chapter 8: Money Power

Exercise 1, page 76

1. 100 pennies
2. 20 nickels
3. 4 quarters
4. 2 50-cent pieces

Exercise 2, page 77

1. $.11
2. $.25
3. $.38
4. $.44
5. $.50
6. $.75
7. $.99
8. $.85

Exercise 3, page 77

1. $5¢ = \$.05$
2. $10¢ = \$.10$
3. $3¢ = \$.03$
4. $80¢ = \$.80$
5. $40¢ = \$.40$
6. $9¢ = \$.09$

Exercise 4, page 78

1. 2 *dimes* + 5 *pennies*
2. 1 *dime* + 9 *pennies*
3. 3 *dimes* + 6 *pennies*
4. 5 *dimes* + 0 *pennies*
5. 7 *dimes* + 2 *pennies*
6. 0 *dimes* + 7 *pennies*
7. 8 *dimes* + 8 *pennies*
8. 9 *dimes* + 1 *penny*

Exercise 5, page 79

1. $3.55
2. $50.70
3. $89.00
4. $153.68

Exercise 6, page 80

1. d
2. b
3. c
4. a

Exercise 7, page 81

1. 20
2. 40
3. 30
4. 50
5. 60
6. 2
7. 3
8. 4
9. 5
10. 1

Exercise 8, page 81

1. $6
2. $3
3. $12
4. $25
5. $75
6. $83

Exercise 9, page 82

1. $.50
2. $.10
3. $.80
4. $.80
5. $1.90
6. $1.20
7. $4.50
8. $5.40
9. $8.20

Exercise 10, page 82

1. $.19
2. $.45
3. $.09
4. $.92
5. $1.47
6. $2.14
7. $3.60
8. $2.51
9. $6.00

Money Checkup, page 83

1. **a.** It separates dollars and cents.
   b. Cents represent *tenths* and *hundredths*.
2. **a.** *hundreds* **b.** *tens* **c.** *ones* **d.** *dimes* **e.** *pennies*
3. **a.** $.35 **b.** $46.08 **c.** $6.00
4. **a.** 10 *dimes* **b.** 10 *pennies*
5. **a.** $5.98 **b.** $12.30 **c.** $18.00

Chapter 9: Solving for Money Amounts

Exercise 1, page 85

1. $534
2. $765
3. $1,200
4. $1,101
5. $722
6. $1,692

Exercise 2, page 86

1. $.48
2. $.63
3. $.55
4. $.90
5. $.87
6. $.63

Exercise 3, page 86

1. $1.17
2. $1.07
3. $1.60
4. $1.03
5. $1.22
6. $1.84
7. $1.48
8. $1.98

Exercise 4, page 87

1. $6.39
2. $7.79
3. $11.88
4. $12.35
5. $8.32
6. $18.04
7. $17.39
8. $39.67
9. $33.10
10. $8.73
11. $11.25
12. $24.61

Exercise 5, page 88

1. $.82
2. $.67
3. $4.44
4. $5.10
5. $5.49
6. $2.51
7. $5.44
8. $4.09
9. $1.27
10. $3.55
11. $3.64
12. $4.29

Exercise 6, page 89

1. $1,696
2. $1,650
3. $762
4. $1,456
5. $1,455
6. $1,196
7. $750
8. $2,340

Exercise 7, page 90

1. $.96
2. $.72
3. $.80
4. $.75
5. $.84
6. $.87
7. $.94
8. $.90

Exercise 8, page 91

1. $1.98
2. $2.25
3. $1.32
4. $4.65
5. $8.20
6. $11.40
7. $46.20
8. $171.80
9. $3.96

Exercise 9, page 91

1. $1.64
2. $2.80
3. $1.45
4. $2.16
5. $6.75
6. $2.97
7. $4.10
8. $29.04
9. $26.20
10. $40.70
11. $30.87
12. $87.42

Exercise 10, page 92

1. $252 + $420 = $672
2. $335 + $670 = $1,005
3. $72 + $360 = $432
4. $1.50 + $7.50 = $9.00
5. $7.20 + $9.00 = $16.20
6. $2.10 + $10.50 = $12.60
7. $13.98 + $69.90 = $83.88
8. $49.00 + $245.00 = $294.00
9. $73.36 + $366.80 = $440.16

Exercise 11, page 93

1. $4.50
2. $4.25
3. $3.60
4. $11.50
5. $9.75
6. $6.50
7. $7.20
8. $9.60

Exercise 12, page 94

1. $2.05
2. $4.52
3. $4.15
4. $3.30
5. $7.89
6. $6.28
7. $9.99
8. $2.22
9. $3.36
10. $3.56
11. $10.75
12. $36.18

Exercise 13, page 95

1. $2.265; round up to $2.27
2. $1.723; round down to $1.72
3. $2.535; round up to $2.54
4. $6.348; round up to $6.35
5. $5.936; round up to $5.94
6. $6.362; round down to $6.36
7. $20.677; round up to $20.68
8. $10.704; round down to $10.70

Exercise 14, page 96

1. a. $1 + $1 + $5 + $2 = $9
   b. $3 + $4 + $3 + $2 = $12
   c. $1 + $5 + $2 + $1 = $9
2. a. 3 × $1 = $3
   b. 2 × $2 = $4
   c. 2 × $5 = $10
3. a. $1 ÷ 4 = $.25
   b. $5 ÷ 2 = $2.50
   c. $2 ÷ 2 = $1

Money Checkup, page 97

1. a. $2 + 2 = $4; $3.24
   b. $3 + $15 = $18; $18.03
   c. $66 − 30 = $36; $35.84
   d. 8 × $7 = $56; $58.80
   e. $10 ÷ 5 = $2; $1.91
   f. $81 ÷ 3 = $27; $26.92
2. a. $14.67
   b. $20.29
   c. $5.33
3. a. 3 × $11.99 = ____; 3 × $12 = $36; $35.97
   b. $20.07 − $11.67 = ____; $20 − $12 = $8; $8.40

Unit 3: Decimals

Chapter 10: Decimal Power

Exercise 1, page 100

1. a. *tenths*
   b. 5
   c. 5
2. a. *tenths*
   b. 9
   c. 1
3. a. *hundredths*
   b. 30
   c. 70
4. a. *hundredths*
   b. 60
   c. 40

Exercise 2, page 101

1. .⑥; *tenths*
2. .①5; *tenths*
3. .⑥; *tenths*
4. .②9; *tenths*
5. .⑤26; *tenths*
6. .0⑧3; *hundredths*
7. .4⑨; *hundredths*
8. .①9; *tenths*

Exercise 3, page 102

1. The denominator is *hundredths*; the number has 2 decimal places.
2. The denominator is *thousandths*; the number has 3 decimal places.
3. The denominator is *tenths*; the number has 1 decimal place.
4. The denominator is *hundredths*; the number has 2 decimal places.
5. The denominator is *tenths*; the number has 1 decimal place.

Exercise 4, page 102

1. .06
2. .004
3. .012
4. .096
5. .008
6. .09

Exercise 5, page 103

1. 2.4
2. 3.36
3. 14.212
4. 26.99

Exercise 6, page 104

1. a. .4 = .40
   b. .60 = .6
2. a. .3 = .30
   b. .70 = .7
3. a. .8 = .80
   b. .20 = .2
4. a. .5 = .50
   b. .50 = .5

Exercise 7, page 105

1. .57, .570
2. .900, .9
3. .824, .82400
4. 3.10, 3.1
5. 7.45, 7.450
6. 5.00, 5
7. 20, 20.00
8. 36.30, 36.3
9. 41.24, 41.240

Exercise 8, page 105

1. a. .80
   b. .50
   c. 1.30
   d. 6.90
   e. 12.60
   f. 22.20
   g. 9.00
   h. 25.00
2. a. .320
   b. .070
   c. .750
   d. 2.180
   e. 7.090
   f. 19.640
   g. 38.000
   h. 50.000

Exercise 9, page 106

1. .36
2. .7
3. .04
4. 3
5. 5.6
6. 10.01
7. 28
8. 2.505

Decimal Checkup, page 107

1. a. The denominator represents the number of equal parts that 1 is divided into.
b. The numerator represents the number of equal parts that are being discussed.
c. Equivalent decimals have the same value.
2. b. *tenths*
c. *hundredths*
a. *thousandths*
3. a. *hundredths*
b. *thousandths*
c. *ten-thousandths*
d. *tenths*
4. a. .250
b. 15.900
c. 3.000
5. a. 4.89
b. .8
c. .07
d. 15

Chapter 11: Solving Decimal Problems

Exercise 1, page 109

1.
```
    . 5 0
  + . 1 7
    . 6 7
```
2.
```
    . 0 9 9
  + . 4 5 0
    . 5 4 9
```
3.
```
    3 . 6 9 4
  + 8 . 1 0 0
   11 . 7 9 4
```
4.
```
    4 . 0 0 0
  +   . 7 5 9
    4 . 7 5 9
```
5.
```
     7 . 6 0
     9 . 0 0
  + 1 2 . 0 9
    2 8 . 6 9
```
6.
```
    . 6 5 3
    . 8 0 0
  + 1 . 4 0 0
    2 . 8 5 3
```

Exercise 2, page 110

1.
```
    . 8 4
  - . 3 0
    . 5 4
```
2.
```
    . 5 0 0
  - . 2 0 9
    . 2 9 1
```
3.
```
    . 7 7 0
  - . 6 1 4
    . 1 5 6
```
4.
```
    3 . 8 5 5
  -   . 7 3 0
    3 . 1 2 5
```
5.
```
    4 . 0 0
  - 1 . 1 6
    2 . 8 4
```
6.
```
    9 . 9 0
  - 7 . 3 5
    2 . 5 5
```
7.
```
   1 0 . 3 0 0
  -  2 . 8 9 8
    7 . 4 0 2
```
8.
```
    2 1 . 5 7
  - 1 6 . 4 0
     5 . 1 7
```

9.
```
    2 0 . 0 0 0
  - 1 5 . 5 5 3
     4 . 4 4 7
```

Exercise 3, page 111

1. 5.0. The problem has 1 decimal place.
2. 1.35. The problem has 2 decimal places.
3. 6.072. The problem has 3 decimal places.
4. 4.488. The problem has 3 decimal places.
5. 2.3000. The problem has 4 decimal places.
6. 1.952. The problem has 3 decimal places.

Exercise 4, page 112

1. 6.30 = 6.3
2. 18.3
3. 3.25
4. 18.96
5. 14.050 = 14.05
6. 32.7
7. 34.4
8. 96.40 = 96.4
9. 11.60 = 11.6

Exercise 5, page 112

1. 4.5
2. 5.6
3. .52
4. 3.21
5. .35
6. .129
7. 1.40 = 1.4
8. 1.80 = 1.8
9. .5030 = .503

Exercise 6, page 113

1. 2.508
2. 17.98
3. 2.175
4. 76.78
5. 244.68
6. 2.2896
7. 7.294
8. 32.40
9. 1.260
10. 8.551

Exercise 7, page 114

1. 2.2
2. .7
3. .038
4. 6.095
5. 12.35
6. 3.383
7. 1.46
8. .856
9. .755
10. 3.067
11. 5.02
12. 2.1175

Exercise 8, page 115

1. 15 $\overline{)\,372}$: $100 \times .15 = 15$ and $100 \times 3.72 = 372.0$ OR 372
2. 3 $\overline{)\,42.5}$: $10 \times .3 = 3$ and $10 \times 4.25 = 42.5$
3. 14 $\overline{)\,172}$: $10 \times 1.4 = 14$ and $10 \times 17.2 = 172$
4. 9 $\overline{)\,81}$: $10 \times .9 = 9$ and $10 \times 8.1 = 81$

Exercise 9, page 116

1. 12
2. 11.375; round to 11.38
3. 2.46
4. 3.21
5. 36
6. 500
7. 50.714; round to 50.71
8. 2
9. 12.5
10. 3
11. 24
12. 433.333; round to 433.33

Exercise 10, page 117

1. $3.500 - 3.128 =$ ____; $4 - 3 =$ about 1 pound
2. $2 \times 3.125 =$ ____; $2 \times 3 =$ about 6 ounces
3. $24.3 \div 6 =$ ____; $24 \div 6 =$ about 4 fluid ounces
4. $1.79 + 2 =$ ____; $2 + 2 =$ about 4 pounds
5. $314.9 \div 5 =$ ____; $300 \div 5 =$ about 60 miles

Decimal Checkup, page 118

1. To make it easier to add, the numbers should have the same denominator as 9.845.
2. Borrow 1 *ten* and regroup as 10 *ones*. Borrow 1 *one* and regroup as 10 *tenths* and add to 6 *tenths*.
3. Yes. The problem has 2 decimal places.
4. $10 \times .5 = 5.0$ or 5 and $10 \times 9.6 = 96.0$ OR 96

Unit 4: Common Fractions

Chapter 12: Fraction Power

Exercise 1, page 121

1. a
2. b
3. c

Exercise 2, page 122

1. a. $\frac{8}{8}$
   b. $\frac{3}{8}$
   c. $\frac{5}{8}$
2. a. $\frac{3}{3}$
   b. $\frac{2}{3}$
   c. $\frac{1}{3}$
3. a. $\frac{5}{5}$
   b. $\frac{3}{5}$
   c. $\frac{2}{5}$
4. a. $\frac{12}{12}$
   b. $\frac{8}{12}$
   c. $\frac{4}{12}$

Exercise 3, page 123

1. $\frac{1}{3}$; thirds are larger equal parts than fourths
2. $\frac{1}{2}$; halves are larger equal parts than fourths
3. $\frac{1}{3}$; thirds are larger equal parts than eighths
4. $\frac{1}{2}$; halves are larger equal parts than fifths
5. $\frac{1}{8}$; eighths are larger equal parts than sixteenths

Exercise 4, page 124

1. $\frac{3}{4}$
2. $\frac{2}{3}$
3. $\frac{3}{5}$
4. $\frac{3}{4}$
5. $\frac{1}{2}$
6. $\frac{9}{12}$

Exercise 5, page 125

1. $\frac{20}{10}$
2. $\frac{16}{16}$
3. $\frac{4}{2}$

Exercise 6, page 125

1. $\frac{7}{4}$
2. $\frac{24}{16}$
3. $\frac{14}{6}$

Exercise 7, page 126

1. a. $1\frac{1}{4}$
   b. $2\frac{3}{4}$
   c. $3\frac{2}{4}$
2. a. $1\frac{4}{8}$
   b. $2\frac{1}{8}$
   c. $2\frac{7}{8}$
   d. $3\frac{3}{8}$
3. a. $1\frac{4}{16}$
   b. $2\frac{8}{16}$
   c. $2\frac{15}{16}$
   d. $3\frac{10}{16}$

Exercise 8, page 126

1. $1\frac{1}{3}$
2. $4\frac{3}{4}$
3. $3\frac{5}{16}$
4. $3\frac{7}{10}$
5. $6\frac{1}{2}$

Exercise 9, page 127

1. 1
2. 3
3. 3
4. 2
5. 4
6. 6
7. 6
8. 7
9. 11
10. 16
11. 21
12. 25

Fraction Checkup, page 128

1. a, b
2. a. $\frac{5}{8}$
   b. $\frac{2}{3}$
   c. $\frac{5}{6}$
3. a. $\frac{1}{8}$
   b. $\frac{6}{8}$
   c. $1\frac{4}{8}$
4. a. $\frac{1}{4}$
   b. $\frac{2}{5}$
   c. $\frac{1}{3}$
   d. $\frac{3}{8}$
5. a. $6\frac{1}{2}$; 7
   b. $2\frac{3}{4}$; 3

Chapter 13: Renaming Fractions

Exercise 1, page 130

1. $\frac{2}{8} = \frac{1}{4}$
2. $\frac{2}{5} = \frac{4}{10}$
3. $\frac{6}{8} = \frac{3}{4}$

Exercise 2, page 130

1. **a.** $\frac{2}{8}$
   b. $\frac{4}{6}$
   c. $\frac{6}{16}$
2. **a.** $\frac{3}{12}$
   b. $\frac{6}{9}$
   c. $\frac{9}{24}$
3. **a.** $\frac{4}{16}$
   b. $\frac{8}{12}$
   c. $\frac{12}{32}$
4. **a.** $\frac{5}{20}$
   b. $\frac{10}{15}$
   c. $\frac{15}{40}$

Exercise 3, page 131

1. $\frac{3}{6}$
2. $\frac{2}{6}$
3. $\frac{6}{8}$
4. $\frac{4}{10}$
5. $\frac{6}{15}$

Exercise 4, page 131

1. 4 or 8
2. 6 or 18
3. 12
4. 15
5. 20
6. 10 or 50
7. 6 or 12
8. 12 or 36

Exercise 5, page 132

1. 3
2. 5
3. 2
4. 2
5. 2, 4, or 8

Exercise 6, page 132

1. $\frac{1}{2}$
2. $\frac{1}{3}$
3. $\frac{2}{3}$
4. $\frac{3}{5}$
5. $\frac{3}{4}$
6. $\frac{2}{3}$
7. $\frac{1}{3}$
8. $\frac{3}{8}$

Exercise 7, page 133

1. $\frac{4}{5}$
2. $\frac{1}{4}$
3. $\frac{3}{5}$
4. $\frac{21}{25}$
5. $\frac{3}{4}$
6. $\frac{1}{8}$

Exercise 8, page 133

1. .25
2. .75
3. .125
4. .33
5. .375
6. .83
7. .6
8. .67

Exercise 9, page 134

1. $1\frac{1}{2} = \frac{3}{2}$
2. $1\frac{6}{8} = \frac{14}{8}$
3. $2\frac{2}{3} = \frac{8}{3}$

Exercise 10, page 134

1. $\frac{17}{4}$
2. $\frac{5}{2}$
3. $\frac{5}{3}$
4. $\frac{7}{4}$
5. $\frac{7}{2}$
6. $\frac{22}{5}$
7. $\frac{7}{3}$
8. $\frac{31}{8}$

Exercise 11, page 135

1. 2
2. 2
3. 1
4. 3
5. $3\frac{1}{2}$
6. $3\frac{1}{3}$
7. $2\frac{1}{4}$
8. $1\frac{3}{5}$

Exercise 12, page 135

1. 2.25
2. 3.5
3. 2.4
4. 1.75
5. 10.5
6. 1.5
7. 1.4
8. 10.33

Fraction Checkup, page 136

1. **a.** $\frac{1}{2} = \frac{2}{4} = \frac{3}{6} = \frac{4}{8} = \frac{5}{10}$
   b. $\frac{1}{3} = \frac{2}{6} = \frac{3}{9} = \frac{4}{12} = \frac{5}{15}$
2. **a.** 15
   b. 8 or 16
3. **a.** $\frac{1}{2}$
   b. $\frac{1}{4}$
   c. $\frac{4}{5}$
4. **a.** $\frac{5}{2}$
   b. $\frac{15}{4}$
   c. $\frac{21}{5}$
5. **a.** $4\frac{1}{2}$
   b. $1\frac{3}{5}$
   c. $4\frac{2}{3}$
6. **a.** .25
   b. .4
   c. 1.375
   d. 5.5

Chapter 14: Fractions and Mixed Numbers

Exercise 1, page 138

1. $\frac{4}{5}$
2. $\frac{11}{12}$
3. $\frac{13}{16}$
4. $\frac{9}{10}$
5. $\frac{6}{8}$
6. $\frac{5}{3}$

Exercise 2, page 138

1. $\frac{3}{4}$
2. $\frac{5}{6}$
3. $\frac{6}{8} = \frac{3}{4}$
4. $\frac{3}{5}$
5. $\frac{2}{2} = 1$
6. $\frac{4}{3} = 1\frac{1}{3}$
7. $\frac{8}{16} = \frac{1}{2}$
8. $\frac{6}{4} = 1\frac{2}{4} = 1\frac{1}{2}$
9. $\frac{3}{2} = 1\frac{1}{2}$
10. $\frac{6}{8} = \frac{3}{4}$
11. $\frac{10}{10} = 1$
12. $\frac{12}{8} = 1\frac{4}{8} = 1\frac{1}{2}$

Exercise 3, page 139

1. $\frac{2}{4} + \frac{3}{4} = \frac{5}{4} = 1\frac{1}{4}$
2. $\frac{4}{6} + \frac{3}{6} = \frac{7}{6} = 1\frac{1}{6}$
3. $\frac{8}{10} + \frac{5}{10} = \frac{13}{10} = 1\frac{3}{10}$
4. $\frac{3}{8} + \frac{2}{8} = \frac{5}{8}$
5. $\frac{4}{10} + \frac{7}{10} = \frac{11}{10} = 1\frac{1}{10}$
6. $\frac{18}{24} + \frac{4}{24} = \frac{22}{24} = \frac{11}{12}$ OR $\frac{9}{12} + \frac{2}{12} = \frac{11}{12}$
7. $\frac{9}{12} + \frac{4}{12} = \frac{13}{12} = 1\frac{1}{12}$
8. $\frac{4}{8} + \frac{5}{8} = \frac{9}{8} = 1\frac{1}{8}$
9. $\frac{5}{16} + \frac{12}{16} = \frac{17}{16} = 1\frac{1}{16}$
10. $\frac{4}{8} + \frac{2}{8} + \frac{3}{8} = \frac{9}{8} = 1\frac{1}{8}$
11. $\frac{4}{12} + \frac{9}{12} + \frac{5}{12} = \frac{18}{12} = 1\frac{6}{12} = 1\frac{1}{2}$
12. $\frac{6}{24} + \frac{8}{24} + \frac{12}{24} = \frac{26}{24} = 1\frac{2}{24} = 1\frac{1}{12}$ OR $\frac{3}{12} + \frac{4}{12} + \frac{6}{12} = \frac{13}{12} = 1\frac{1}{12}$

Exercise 4, page 140

1. $5\frac{2}{4} = 5\frac{1}{2}$
2. $6\frac{4}{8} = 6\frac{1}{2}$
3. $4\frac{8}{10} = 4\frac{4}{5}$
4. $4\frac{3}{4}$
5. $4\frac{5}{6}$
6. $7\frac{3}{4}$
7. $6\frac{8}{8} = 7$
8. $7\frac{2}{2} = 8$
9. $7\frac{4}{4} = 8$
10. $6\frac{1}{8}$
11. $8\frac{1}{4}$
12. $8\frac{1}{6}$

Exercise 5, page 141

1. $\frac{1}{3}$
2. $\frac{3}{8}$
3. $\frac{7}{12}$
4. $\frac{6}{16} = \frac{3}{8}$
5. $\frac{4}{10} = \frac{2}{5}$

Exercise 6, page 141

1. $\frac{1}{4}$
2. $\frac{1}{6}$
3. $\frac{5}{10} = \frac{1}{2}$
4. $\frac{1}{6}$
5. $\frac{3}{8}$
6. $\frac{7}{15}$
7. $\frac{9}{16}$
8. $\frac{1}{12}$

Exercise 7, page 142

1. $\frac{3}{3}$
2. $\frac{2}{2}$
3. $\frac{5}{5}$
4. $\frac{10}{10}$
5. $\frac{8}{8}$
6. $\frac{16}{16}$
7. $\frac{4}{4}$
8. $\frac{6}{6}$

Exercise 8, page 142

1. $\frac{1}{3}$
2. $\frac{1}{4}$
3. $1\frac{5}{8}$
4. $2\frac{1}{5}$
5. $4\frac{1}{4}$
6. $2\frac{1}{10}$
7. $8\frac{1}{2}$
8. $6\frac{1}{3}$
9. $7\frac{1}{2}$

Exercise 9, page 143

1. $1\frac{1}{3}$
2. $2\frac{1}{2}$
3. $\frac{1}{5}$
4. $5\frac{2}{4} - 1\frac{1}{4} = 4\frac{1}{4}$
5. $4\frac{4}{8} - 1\frac{3}{8} = 3\frac{1}{8}$
6. $5\frac{3}{6} - 2\frac{2}{6} = 3\frac{1}{6}$
7. $7\frac{5}{4} - 3\frac{3}{4} = 4\frac{2}{4} = 4\frac{1}{2}$
8. $6\frac{6}{5} - 4\frac{3}{5} = 2\frac{3}{5}$
9. $6\frac{3}{6} - 3\frac{4}{6} = ?;\ 5\frac{9}{6} - 3\frac{4}{6} = 2\frac{5}{6}$
10. $2\frac{2}{2} - 2\frac{1}{2} = \frac{1}{2}$
11. $7\frac{7}{8} - 5\frac{5}{8} = 2\frac{2}{8}$
12. $9\frac{1}{4} - 8\frac{2}{4} = ?;\ 8\frac{5}{4} - 8\frac{2}{4} = \frac{3}{4}$

Exercise 10, page 144

1. $\frac{1}{2} \times \frac{3}{4} = \frac{3}{8}$
2. $\frac{1}{4} \times \frac{1}{2} = \frac{1}{8}$
3. $\frac{1}{3} \times \frac{9}{10} = \frac{9}{30} = \frac{3}{10}$
4. $\frac{1}{2} \times \frac{1}{2} = \frac{1}{4}$
5. $\frac{1}{4} \times \frac{1}{4} = \frac{1}{16}$
6. $\frac{1}{3} \times \frac{3}{4} = \frac{3}{12} = \frac{1}{4}$

Exercise 11, page 145

1. $\frac{8}{2} = 4$
2. $\frac{12}{2} = 6$
3. $\frac{9}{2} = 4\frac{1}{2}$
4. $\frac{15}{2} = 7\frac{1}{2}$
5. $\frac{5}{4} = 1\frac{1}{4}$
6. $\frac{9}{3} = 3$
7. $\frac{12}{5} = 2\frac{2}{5}$
8. $\frac{8}{10} = \frac{4}{5}$
9. $\frac{21}{4} = 5\frac{1}{4}$
10. $\frac{30}{3} = 10$
11. $\frac{36}{8} = 4\frac{4}{8} = 4\frac{1}{2}$
12. $\frac{20}{5} = 4$

Exercise 12, page 146

1. $\frac{1}{2} \times \frac{3}{2} = \frac{3}{4}$
2. $\frac{1}{3} \times \frac{7}{2} = \frac{7}{6} = 1\frac{1}{6}$
3. $\frac{1}{4} \times \frac{19}{4} = \frac{19}{16} = 1\frac{3}{16}$
4. $\frac{3}{4} \times \frac{11}{2} = \frac{33}{8} = 4\frac{1}{8}$
5. $\frac{2}{3} \times \frac{19}{3} = \frac{38}{9} = 4\frac{2}{9}$
6. $\frac{1}{2} \times \frac{29}{4} = \frac{29}{8} = 3\frac{5}{8}$
7. $\frac{1}{3} \times \frac{21}{4} = \frac{21}{12} = 1\frac{9}{12} = 1\frac{3}{4}$
8. $\frac{1}{2} \times \frac{27}{4} = \frac{27}{8} = 3\frac{3}{8}$
9. $\frac{3}{8} \times \frac{5}{2} = \frac{15}{16}$

Exercise 13, page 147

1. $\frac{2}{1}$
2. $\frac{4}{3}$
3. $\frac{8}{5}$
4. $\frac{10}{3}$
5. $\frac{16}{7}$

Exercise 14, page 147

1. $\frac{9}{1} \times \frac{3}{1} = 27$
2. $\frac{14}{3} \times \frac{2}{1} = \frac{28}{3} = 9\frac{1}{3}$
3. $\frac{17}{2} \times \frac{2}{1} = \frac{34}{2} = 17$
4. $\frac{27}{4} \times \frac{4}{3} = \frac{108}{12} = 9$
5. $\frac{27}{4} \times \frac{3}{2} = \frac{81}{8} = 10\frac{1}{8}$
6. $\frac{10}{1} \times \frac{2}{1} = 20$
7. $\frac{22}{3} \times \frac{2}{1} = \frac{44}{3} = 14\frac{2}{3}$
8. $\frac{7}{1} \times \frac{4}{3} = \frac{28}{3} = 9\frac{1}{3}$

Exercise 15, page 148

1. $3\frac{1}{2} = \frac{7}{2};\ \frac{2}{7}$
2. $4\frac{2}{3} = \frac{14}{3};\ \frac{3}{14}$
3. $1\frac{5}{8} = \frac{13}{8};\ \frac{8}{13}$
4. $8 = \frac{8}{1};\ \frac{1}{8}$
5. $6\frac{1}{4} = \frac{25}{4};\ \frac{4}{25}$
6. $15 = \frac{15}{1};\ \frac{1}{15}$

Exercise 16, page 148

1. $\frac{5}{2} \times \frac{1}{2} = \frac{5}{4} = 1\frac{1}{4}$
2. $\frac{7}{4} \times \frac{1}{4} = \frac{7}{16}$
3. $\frac{14}{3} \times \frac{3}{4} = \frac{42}{12} = 3\frac{6}{12} = 3\frac{1}{2}$
4. $\frac{13}{2} \times \frac{2}{3} = \frac{26}{6} = 4\frac{2}{6} = 4\frac{1}{3}$
5. $\frac{3}{4} \times \frac{1}{2} = \frac{3}{8}$
6. $\frac{1}{2} \times \frac{1}{2} = \frac{1}{4}$
7. $\frac{35}{4} \times \frac{4}{5} = \frac{140}{20} = 7$
8. $\frac{2}{3} \times \frac{3}{4} = \frac{6}{12} = \frac{1}{2}$

Fraction Checkup, page 149

1. a. addition
   b. subtraction
   c. addition
2. a. $\frac{6}{8} = \frac{3}{4}$
   b. $1\frac{4}{15}$
   c. $4\frac{1}{4}$
   d. $\frac{3}{4}$
   e. $\frac{1}{8}$

   f. $3\frac{2}{3}$
   g. $\frac{3}{8}$
   h. $1\frac{1}{4}$
   i. 4
   j. $1\frac{1}{2}$
   k. 10
   l. $1\frac{5}{8}$

Unit 5: Ratios and Percents

Chapter 15: Ratio Power

Exercise 1, page 152
1. 15 to 1   **3.** 8 to 1
2. 20 to 1   **4.** 37 to 1

Exercise 2, page 152
1. 25 to 100   **4.** 3 to 4
2. 75 to 100   **5.** 4 to 5
3. 1 to 4   **6.** 8 to 10

Exercise 3, page 153
1. $\frac{2}{1}$ **7.** $\frac{8}{5}$
2. $\frac{3}{4}$ **8.** $\frac{1}{18}$
3. $\frac{5}{10}$ **9.** $\frac{4}{3}$
4. $\frac{1}{2}$ **10.** $\frac{5}{8}$
5. $\frac{3}{6}$ **11.** $\frac{18}{1}$
6. $\frac{6}{3}$ **12.** $\frac{10}{5}$

Exercise 4, page 154
1. $\frac{1}{2}$ **6.** $\frac{1}{2}$
2. $\frac{2}{3}$ **7.** $\frac{3}{4}$
3. $\frac{3}{1}$ **8.** $\frac{7}{6}$
4. $\frac{2}{1}$ **9.** $\frac{3}{1}$
5. $\frac{1}{2}$ **10.** $\frac{6}{5}$

Exercise 5, page 154
1. $\frac{4}{2} = \frac{2}{1}$ **3.** $\frac{180}{12} = \frac{15}{1}$
2. $\frac{15}{25} = \frac{3}{5}$ **4.** $\frac{160}{200} = \frac{4}{5}$

Exercise 6, page 155
1. $\frac{2}{4} = \frac{1}{2}$ **4.** $\frac{14}{10} = \frac{7}{5}$
2. $\frac{6}{2} = \frac{3}{1}$ **5.** $\frac{25}{15} = \frac{5}{3}$
3. $\frac{10}{20} = \frac{1}{2}$ **6.** $\frac{32}{24} = \frac{4}{3}$

Exercise 7, page 156
Cross products can be in any order:
1. a. $2 \times 12 = 24$ **3. a.** $1 \times 120 = 120$
   b. $8 \times 3 = 24$ **b.** $20 \times 6 = 120$
2. a. $1 \times 24 = 24$ **4. a.** $7 \times 100 = 700$
   b. $6 \times 4 = 24$ **b.** $70 \times 10 = 700$

Exercise 8, page 156
1. True: $1 \times 24 = 24$; $12 \times 2 = 24$
2. True: $1 \times 8 = 8$; $2 \times 4 = 8$
3. False: $1 \times 12 = 12$; $5 \times 4 = 20$
4. True: $2 \times 12 = 24$; $8 \times 3 = 24$
5. True: $3 \times 15 = 45$; $9 \times 5 = 45$
6. False: $3 \times 20 = 60$; $12 \times 4 = 48$
7. True: $1 \times 30 = 30$; $10 \times 3 = 30$
8. False: $1 \times 120 = 120$; $40 \times 2 = 80$
9. False: $3 \times 16 = 48$; $7 \times 8 = 56$
10. True: $1 \times 72 = 72$; $36 \times 2 = 72$

Exercise 9, page 157
1. $u = 9$: $3 \times 15 = 45$; $45 \div 5 = 9$
2. $s = 8$: $4 \times 2 = 8$; $8 \div 1 = 8$
3. $x = 1$: $9 \times 3 = 27$; $27 \div 27 = 1$
4. $g = 2$: $6 \times 5 = 30$; $30 \div 15 = 2$
5. $a = 5$: $4 \times 20 = 80$; $80 \div 16 = 5$
6. $r = 3$: $1.5 \times 6 = 9$; $9 \div 3 = 3$

Exercise 10, page 157
1. a. $\frac{15}{1} = \frac{x}{4.5}$
   b. $x = \$67.50$: $15 \times 4.5 = 67.5$; $67.5 \div 1 = 67.5$
2. a. $\frac{225}{10} = \frac{x}{1}$
   b. $x = 22.5$ miles per gallon: $225 \times 1 = 225$;
      $225 \div 10 = 22.5$

Ratio Checkup, page 158
1. c: The two amounts are compared for their difference by subtraction.
2. a. $\frac{3}{5}$
   b. $\frac{15}{1}$
   c. $\frac{150}{2} = \frac{75}{1}$
3. a. Yes: $2 \times 9 = 18$; $6 \times 3 = 18$
   b. Yes: $4 \times 25 = 100$; $5 \times 20 = 100$
   c. Yes: $7 \times 56 = 392$; $49 \times 8 = 392$
4. a. $x = 540$ miles: $\frac{90}{1} = \frac{x}{6}$; $90 \times 6 = 540$;
     $540 \div 1 = 540$
   b. $x = \$44.50$: $\frac{2}{89} = \frac{1}{x}$; $1 \times 89 = 89$; $89 \div 2 = \$44.5$

Chapter 16: Percent Power

Exercise 1, page 160

1. a. $\frac{7}{10} = \frac{70}{100}$
   70 percent is shaded.

b. $\frac{3}{10} = \frac{30}{100}$
   30 percent is unshaded.

2. a. $\frac{2}{10} = \frac{20}{100}$
   20 percent is shaded.

b. $\frac{8}{10} = \frac{80}{100}$
   80 percent is unshaded.

Exercise 2, page 160

1. 25 percent = 25%
2. 15 percent = 15%
3. 7 percent = 7%
4. 18 percent = 18%
5. Forty-two percent = 42%

Exercise 3, page 161

1. a. 60 questions
   b. 15 math questions
   c. 45 reading questions
2. a. $60
   b. $9
   c. $51

3. a. 20 coins
   b. 4 dimes
   c. 6 pennies
   d. 10 nickels
4. a. 200 invitations
   b. 96 yes
   c. 44 no
   d. 60 no reply

Exercise 4, page 162

1. $\frac{10}{100} = \frac{1}{10}$
2. $\frac{5}{100} = \frac{1}{20}$
3. $\frac{42}{100} = \frac{21}{50}$
4. $\frac{25}{100} = \frac{1}{4}$
5. $\frac{75}{100} = \frac{3}{4}$
6. $\frac{80}{100} = \frac{4}{5}$
7. $\frac{60}{100} = \frac{3}{5}$
8. $\frac{93}{100}$

Exercise 5, page 162

1. $\frac{20}{100} = \frac{\$69.80}{\$349}$
2. $\frac{5}{100} = \frac{\$1.25}{\$25.00}$
3. $\frac{50}{100} = \frac{13}{26}$
4. $\frac{30}{100} = \frac{72}{240}$

Exercise 6, page 163

1. $\frac{25}{100} = \frac{?}{20}$; ? = 5
2. $\frac{15}{100} = \frac{?}{20}$; ? = 3
3. $\frac{20}{100} = \frac{?}{20}$; ? = 4
4. $\frac{30}{100} = \frac{?}{20}$; ? = 6
5. $\frac{10}{100} = \frac{?}{20}$; ? = 2

Exercise 7, page 164

1. .10 = .1
2. .05
3. .42
4. .25
5. .75
6. .80 = .8

Exercise 8, page 164

1. $\frac{10}{100} = \frac{1}{10}$
2. $\frac{5}{100} = \frac{1}{20}$
3. $\frac{42}{100} = \frac{21}{50}$
4. $\frac{25}{100} = \frac{1}{4}$
5. $\frac{75}{100} = \frac{3}{4}$
6. $\frac{80}{100} = \frac{4}{5}$

Exercise 9, page 165

1. a. $.70 \times 120 = $ ____
   b. $\frac{7}{10} \times 120 = $ ____
2. a. $.15 \times 28 = $ ____
   b. $\frac{3}{20} \times 28 = $ ____
3. a. $.20 \times 84 = $ ____
   b. $\frac{1}{5} \times 84 = $ ____

4. a. $.10 \times 2,500 = $ ____
   b. $\frac{1}{10} \times 2,500 = $ ____
5. a. $.60 \times 258 = $ ____
   b. $\frac{3}{5} \times 258 = $ ____
6. a. $.30 \times 2,800 = $ ____
   b. $\frac{3}{10} \times 2,800 = $ ____

Percent Checkup, page 166

1. 25%
2. $\frac{12}{100}$; 12%; .12
3. 75 to 100 OR $\frac{75}{100}$
4. a. .60 = .6
   b. $\frac{60}{100} = \frac{3}{5}$
5. $1,200
6. a. 30%
   b. 20%
   c. 5%
7. a. $\frac{25}{100} = \frac{?}{1,200}$; $.25 \times 1,200 = $ ____; $\frac{1}{4} \times 1,200 = $ ____
   b. $\frac{5}{100} = \frac{?}{1,200}$; $.05 \times 1,200 = $ ____; $\frac{1}{20} \times 1,200 = $ ____